AF412786

Natural History and Evolution of *Cryptopecten*
(A Cenozoic-recent Pectinid Genus)

1a
1b
2
3a
3b
4a
4b
5
6
7
8
9
10

Frontispiece

(All figures ×2)

Cryptopecten bullatus (Dautzenberg and Bavay)

    Fig. 1a, b.   Living specimen with well developed gonad (UMUT RM16184a); Sample *Sr (2)* *(B)* collected on June 15, 1983.

Cryptopecten vesiculosus (Dunker)

    Fig. 2.   Living specimen with poorly developed gonad (UMUT RM16186a); Sample *Sr* collected on June 15, 1983. Phenotype Q.

    Fig. 3a, b.   Fresh conjoined valves (UMUT RM16138d); Sample *Is*. Phenotype *R*.

    Fig. 4a, b.   Fresh conjoined valves (UMUT RM16138g); Sample *Is*. Phenotype Q.

    Figs. 5–9.   Living specimens with well developed gonad (Sample *Jg (24)*–*Jg (26)*, collected on August 3, 1983 at the same station. Fig. 5: Phenotype Q (UMUT RM16191a), Fig. 6: Phenotype *R* (UMUT RM16192a), Fig. 7: Phenotype *R* (UMUT RM16191b), Fig. 8: Phenotype Q (UMUT RM16191c), Fig. 9: Phenotype Q (UMUT RM16193a), Fig. 10: Phenotype *R* (UMUT RM16192b).

NATURAL HISTORY AND EVOLUTION OF *CRYPTOPECTEN* (A CENOZOIC-RECENT PECTINID GENUS)

Itaru Hayami

*Department of Historical Geology and Palaeontology,
The University Museum, The University of Tokyo*

UNIVERSITY OF TOKYO PRESS

The University Museum, The University of Tokyo, Bulletin No. 24, 1984

CONTENTS

Chapter 1   Introduction 1

Chapter 2   Material and Method of Study 4

Chapter 3   Geographic Distribution and Habitat of Living *Cryptopecten* 8

Chapter 4   Stratigraphic Distribution and Mode of Occurrence of
Fossil *Cryptopecten* 16

Chapter 5   Ontogenetical Development of Surface Sculpture in *Cryptopecten* 21

Chapter 6   Reproductive Season and Growth Rings of *Cryptopecten* 24

Chapter 7   Intrapopulational Variation and Relative Growth 30
  A.   Continuous Variation 30
    1)   Shell Size and Growth Rate 34
    2)   Length versus Height 36
    3)   Length versus Thickness 41
    4)   Length versus Thickness of Conjoined Valves 45
    5)   Length versus Dorsal Length of Wings 46
    6.   Number of Radial Ribs 47
  B.   Discontinuous Variation 51
    1)   Color Polymorphism 52
    2)   Dimorphism of Surface Sculpture in *C. vesiculosus* 52
  C.   Evaluation of Ecophenotypic Effect 60

Chapter 8   Geographic Variation of *Cryptopecten vesiculosus* 62
  1)   Shell Size and Growth Rate 62
  2)   Shape of Shell (Bivariate morphometric characters) 63
  3)   Number of Radial Ribs 65
  4)   Coloration of Shell 65
  5)   Surface Sculpture (Phenotypic frequency) 69
  6)   Causal Evaluation of Geographic Variation 69

Chapter 9   Phyletic Evolution of *Cryptopecten vesiculosus* 71
  1)   Shell Size and Growth Rate 71
  2)   Shape of Shell 72
  3)   Number of Radial Ribs 73
  4)   Surface Sculpture (Phenotypic Frequency) 73

Chapter 10   Inferred Speciation of *Cryptopecten* 78

Chapter 11   Discussions on Some General Problems Regarding Patterns of
Morphological Change 83

Chapter 12   Systematic Description of *Cryptopecten* 88
  Genus *Cryptopecten* Dall, Bartsch and Rehder, 1938 88
  Cryptopecten bullatus (Dautzenberg and Bavay) 96
  Cryptopecten nux (Reeve) 99
  Cryptopecten nux nux (Reeve) 100
  Cryptopecten nux sematensis (Oyama) 103
  Cryptopecten vesiculosus (Dunker) 104

  Cryptopecten vesiculosus vesiculosus (Dunker) 105
  Cryptopecten vesiculosus makiyamai subsp. nov. 109
  Cryptopecten phrygium (Dall) 110
  Cryptopecten spinosus sp. nov. 111
  Cryptopecten yanagawaensis (Nomura and Zinbo) 113
Chapter 13 Summary and Conclusions 116
 1) Systematics and Distribution 116
 2) Reproduction and Growth Rings 117
 3) Continuous Variation and Relative Growth 117
 4) Sculpture Dimorphism 118
 5) Geographic Variation 118
 6) Phyletic Evolution 119
 7) Speciation and Extinction 119
 8) Significance of Phenotypic Substitution in Macroevolution 120
 9) Further Problems 121
List of Examined Samples 122
Acknowledgments 139
References 141
Postscript 149
Plates

LIST OF TEXT-FIGURES

Fig. 1. Map showing the localities of Recent samples of *Cryptopecten* around
the southwestern part of the Japanese Islands 10
Fig. 2 Map showing the distribution of four extant species of *Cryptopecten* 12
Fig. 3 Bathymetric distribution of *Cryptopecten* in the western Pacific 14
Fig. 4 Map showing the localities of studied fossil samples of *Cryptopecten*
in the Japanese Islands 18
Fig. 5 Stratigraphic position of *Cryptopecten* samples in the Plio-Pleistocene
sequence of Boso Peninsula near Tokyo 20
Fig. 6 Subvertical sections of some adult specimens of *Cryptopecten vesiculosus*,
showing significant stepwise growth rings 25
Fig. 7 Logistic curves of absolute growth inferred from the growth rings in some
selected individuals belonging to *Cryptopecten vesiculosus* 26
Fig. 8 Internal morphology of *Cryptopecten vesiculosus*, showing linear
measurements 31
Fig. 9 Patterns of absolute growth in two phenotypes of *Cryptopecten vesiculosus*,
inferred from the shell height from the origin of growth to each stepwise
growth ring 33
Fig. 10 Example of the frequency distribution of H_1 (shell height
at the first growth ring). Sample *Is* 34
Fig. 11. Example of the frequency distribution of H_1 (shell height
at the first growth ring). Sample *Ms* 35
Fig. 12. Example of the frequency distribution of H_1 (shell height
at the first growth ring). Sample *Jz* 36
Fig. 13. Example of the allometric relation between L (length) and H
(height). Sample *Jg* 39
Fig. 14. Example of the allometric relation between L (length) and T
(thickness of odd valve). Sample *Jg* 41
Fig. 15. Reduced major axes showing allometric relation between L (length) and T
(thickness of right valve) in some selected samples of *Cryptopecten*.................. 42
Fig. 16. Reduced major axes showing allometric relation between L (length) and T
(thickness of left valve) in some selected samples of *Cryptopecten* 43
Fig. 17. Example of the allometric relation between L (length) and D
(length of dorsal margin). Sample *Jg* 47
Fig. 18. Frequency distribution of the prominence of radial ribs in the sample *Is*
of *Cryptopecten vesiculosus* 53
Fig. 19. Frequency distribution of the prominence of radial ribs in the sample *Jz*
of *Cryptopecten vesiculosus* 54
Fig. 20. Geographic variation of the average number of radial ribs in selected
Recent samples of *Cryptopecten vesiculosus* 64
Fig. 21. Geographic variation of the phenotypic frequency in selected Recent
samples of *Cryptopecten vesiculosus* 66
Fig. 22. Chronological change of the average number of radial ribs in the stock
of *Cryptopecten vesiculosus* 72
Fig. 23. Chronological change of phenotypic frequency in the stock
of *Cryptopecten vesiculosus* 74

Fig. 24.   Inferred interspecific relation of *Cryptopecten* species 79
Fig. 25.   Sculpture dimorphism of *Aequipecten commutatus* 92
Fig. 26.   Generalized diagram of biogeography (longitudinal distribution)
of *Cryptopecten* through geologic time 94

LIST OF TABLES

Table 1.   Examined dated living specimens of *Cryptopecten vesiculosus* 5
Table 2.   Examined dated living specimens of *Cryptopecten bullatus* 9
Table 3.   Average shell height from hinge-line to each growth ring,
presumably indicating annual shell growth 34
Table 4.   Allometric relation between L and H (Right valves) 37
Table 5.   Allometric relation between L and H (Left valves) 38
Table 6.   Allometric relation between L and C (Conjoined valves) 40
Table 7.   Allometric relation between L and T (Right valves) 40
Table 8.   Allometric relation between L and T (Left valves) 44
Table 9.   Allometric relation between L and D (Right valves) 45
Table 10.   Allometric relation between L and D (Left valves) 46
Table 11.   Variation of the number of radial ribs in selected population samples 48
Table 12.   Phenotypic frequency in each sample of *Cryptopecten vesiculosus* 55
Table 13.   Maximum shell size and average growth amount in some selected
samples of *Cryptopecten vesiculosus* 63
Table 14.   Results of F test and t test indicating the significance for the differences
of variance and mean of number of radial ribs between Recent samples 65
Table 15.   Chi-square matrix indicating the significance for the difference of
phenotypic frequency between Recent samples (2×2 contingency tables) 68

LIST OF PLATES

Frontispiece.   Living specimens of *Cryptopecten vesiculosus* and *Cryptopecten bullatus*
Plate 1.   *Cryptopecten bullatus* (Dautzenberg and Bavay)
Plate 2.   *Cryptopecten bullatus* (Dautzenberg and Bavay) and *Cryptopecten nux nux* (Reeve)
Plate 3.   *Cryptopecten nux nux* (Reeve), *Cryptopecten nux sematensis* (Oyama) and *Cryptopecten
vesiculosus vesiculosus* (Dunker)
Plate 4.   *Cryptopecten vesiculosus vesiculosus* (Dunker)
Plate 5.   *Cryptopecten vesiculosus vesiculosus* (Dunker)
Plate 6.   *Cryptopecten vesiculosus vesiculosus* (Dunker)
Plate 7.   *Cryptopecten vesiculosus vesiculosus* (Dunker) and *Cryptopecten vesiculosus makiyamai*
subsp. nov.
Plate 8.   *Cryptopecten spinosus* sp. nov. and *Cryptopecten yanagawaensis* (Nomura and Zinbo)
Plate 9.   *Cryptopecten bullatus* (Dautzenberg and Bavay), *Cryptopecten nux nux* (Reeve) and
Cryptopecten phrygium (Dall)
Plate 10.   *Cryptopecten vesiculosus vesiculosus* (Dunker) and *Cryptopecten bullatus* (Dautzenberg
and Bavay) (SEM photographs)
Plate 11.   *Cryptopecten vesiculosus vesiculosus* (Dunker) and *Cryptopecten bullatus* (Dautzenberg
and Bavay) (SEM photographs)

Plate 12. *Cryptopecten nux nux* (Reeve), *Cryptopecten spinosus* sp. nov. and *Cryptopecten vesiculosus vesiculosus* (Dunker) (SEM photographs)
Plate 13. *Cryptopecten vesiculosus vesiculosus* (Dunker) (SEM photographs)

CHAPTER   1

Introduction

In recent years lively argument has arisen about such general problems of evolutionary paleontology as punctuated patterns of evolution (Eldredge and Gould, 1972; Gould and Eldredge, 1977), higher-level selection acting upon populations and species (Stanley, 1979; Gould, 1982), and the role of random effects in apparently directional morphological changes (Raup, Gould, Schopf and Simberloff, 1973; Raup, 1977). Punctuationists seem to hold that macroevolution as illustrated by fossil records is not explained by gradualistic morphological change within a lineage or by an easy expansion of the acknowledged mechanisms of microevolution.

In the old days the science of paleontology was developed largely by inductive and empirical methods. *Graphy* almost always preceded *logy*. Recent rapid development in various branches of biological and earth sciences as well as revolutionary advances in techniques, however, has given much stimulation to the methodology of evolutionary paleontology. Deductive and theoretical approaches have become more common. Good models and thoughtful extrapolations from accepted opinions in neontology are, of course, quite useful, but descriptive studies of fossil records on the basis of sound concepts are equally important. Without good data on real populations, statistical treatment of morphology would be carried out in vain, and models or hypotheses on the general pattern of evolutionary change would not be able to be tested. In paleontology, as in any other branch of natural history, induction and deduction (or *graphy* and *logy*) are like the two wheels of a cart; both are necessary if the field is to advance.

In Japan, Quaternary marine molluscan fossils, in spite of their abundance owing to rapid crustal uplift in many local areas, have rarely been treated as material for the study of evolutionary process. The reason is, I presume, chiefly because any morphological change that may be present within this period was considered to be too insignificant. More than 70 percent of Early Pleistocene molluscs are actually referable to living species (Stanley, Addicott and Chinzei, 1979). This very fact however gives Quaternary fossils a great advantage for evolutionary studies, because we may understand detailed phyletic evolution and, with good luck, the process of speciation on the basis of well-preserved material and various results of neontological investigations on living populations. Even if genetic experiment is technically difficult, it is still possible to carry out considerable causal evaluation of morphological variation with various kinds of circumstantial evidence. These considerations have prompted me to study the evolutionary pattern and process of the pectinid genus *Cryptopecten*.

In 1970 I happened to find that very striking discontinuous variation of surface sculpture exists in some fossil and Recent samples of this scallop. I immediately remembered Kurtén's (1955) impressive paper on the dimorphism and evolution in the first upper molar of Pleistocene—Recent bears which I had read several years before. The industrial melanism of *Biston betularia*, a famous example of natural selection (Kettlewell,

2

1961; Ford, 1964), and Mayr's (1963, 1969) discussions on individual variation as well as Falconer's (1960) and Kimura's (1960) textbooks on population genetics were also important sources of my knowledge and interest. After two years of field and laboratory works, I prepared manuscripts on the evolutionary change of *C. vesiculosus* on the basis of 12 fossil and 5 Recent samples (Hayami, 1972, 1973). In these papers the change of phenotypic frequency after the Middle Pleistocene was tentatively interpreted as resulting from the accumulation of a mutant gene by natural selection, and it was emphasized that intensive studies on such discontinuous variations in lineages with living end members, if adequately combined in the future with genetic experiments, would contribute to the development of evolutionary theory.

The genetic background of this discontinuous variation is still obscure, but phenotypic substitution was recognized as one of the possible causes for the punctuational morphological changes often observed in fossil records, as was generalized by Hayami and Ozawa (1975). Subsequently some authors discussed general problems of evolutionary paleontology and biostratigraphy citing our discussion on *C. vesiculosus* (Reyment, 1975, 1980; Eldredge and Gould, 1977; Gould and Eldredge, 1977; Raup, 1977; Stanley *in* Fairbridge and Jablonski, 1979).

Cryptopecten mainly consists of lower sublittoral to bathyal species, and is not necessarily ideal material for the study of "paleogenetics", because direct observation of the ecology and embryology at its habitat and Mendelian experiments seem to be considerably difficult. Some intertidal, fresh-water, and land molluscs are more advantageous for ecological and genetic studies. As enumerated before (Hayami, 1972), materials that satisfy the following conditions are preferable for studies of this kind: (1) abundance both in living and fossil states, (2) readiness of culture, (3) readiness of observations on general biology (especially, mode of life, growth and reproduction, population structure and functional significance of morphology), (4) short generation (necessary condition for Mendelian experiments), (5) relatively simple evolutionary lineage(s), (6) presence of genetic polymorphism preferably recognized in the morphology of hard tissue, (7) nearly panmictic populations (without a wide range of geographic variation in phenotypic frequency) and (8) inferable absolute age for each fossil sample. My own observations of extant gastropods and bivalves, however, has shown that materials satisfying many of these conditions are rather rare.

Land snails are certainly advantageous with respect to readiness of culture and ecological observation. As exemplified by Gould's (1966) surprising work on *Poecilozonites* in the Quaternary of Bermuda, studies assuming "an evolutionary microcosm" may be possible on this material in some oceanic islands (e. g., some species of *Mandarina* in the Bonin Islands). Generally speaking, however, populations of land snails are highly localized, and the phenotypic frequency often changes drastically within a small area (Lamotte, 1951; Sheppard, 1952; Komai and Emura, 1955). Diver's (1929) pioneer study on genetic variation in fossil populations of *Cepaea* notwithstanding, evolutionary change is hard to detect owing to the wide range of geographic variation. Moreover, in Japan (except for some subtropical islands) good fossil samples of land snails are rather rare, probably because of soil and climate conditions unfavorable for preservation.

Gastropods and bivalves in embayments and intertidal waters may be more tolerant of conditions in aquaria than those in oceanic waters. Some are considerably well represented in fossils. As the result of Yoda's and my preliminary observations (unpublished data) on three ubiquitous species of *Batillaria* (potamidid gastropods) in Japan, however, the relative frequency of white-banded individuals was found to vary greatly from one embayment to another, suggesting relatively weak gene flow between local populations in spite of their free-swimming larval stage. According to Colton (1922), Moore (1936) and Berry and Crothers (1974), there is also a significantly wide geographic variation in *Nucella lapillus*, a muricid gastropod, common in the intertidal zone around Great Britain. Such non-panmictic species may be useful for examining the influence of environmental factors and differential selection pressure among different areas, but they are generally unsuitable for the study of evolutionary process because the net amount of chronological change may be difficult to detect.

Color polymorphism is a widespread phenomenon in various groups of gastropods and bivalves. Though the variability of ground color is generally difficult to recognize in fossils, color pattern (band, stripe, spot, etc.) is often observable not only in Quaternary but also in much older molluscs, particularly in gastropods with compact shells. However, the recognition of color polymorphism in fossils, even if more clearly observable under ultraviolet light (e. g., Wilson, 1975), depends largely upon a favorable state of preservation. Thus for practical reasons, polymorphism in the sculpture of shell, if it is controlled by genetic cause, is more informative.

Phenotypic discrimination in Recent and fossil samples of *C. vesiculosus* is so easy that one can quickly recognize the phenotypic frequency of a large population sample almost regardless of the state of preservation, differential sorting during post-mortem transportation, incompleteness and deformation of specimens, and age heterogeneity. Such clean-cut discontinuous variation and abundant occurrence both in fossils and present seas are rarely found in other molluscan species. Recent great progress in the chronology and correlation of the Neogene and Quaternary sediments in Japan has made possible the estimation of the absolute ages of many fossil beds yielding *Cryptopecten* samples. These factors lead me to return to a more exhaustive study on the systematics and evolution of *Cryptopecten*.

The Pectinidae are a characteristic bivalve family which has flourished mainly on sublittoral sandy substrates since the Triassic. Their tests are generally tough and well preserved in sediments owing to the stable foliated shell structure. Strongly abraded valves are often found in fossil beds, but they are commonly free from fragmentation except for the auricular parts. Large living and fossil samples are frequently obtainable in some species. Various quantitative characters are easily counted or measured owing to the strong and regular surface ornamentation and easily defined orientation. The ecology, physiology, anatomy and reproduction of several living species have been intensively studied by a number of malacologists (e. g., Dakin, 1909) and workers concerned with commercial fisheries and cultivation (e. g., Yoshida, 1964). The functional significance of shell morphology in some pectinids has been successfully interpreted by several malacologists and paleobiologists (Yonge, 1936; Stanley, 1970; Gould, 1971; Thayer, 1972; Waller, 1972a, b). Though the genetic variation and geographic variation of Indo-Pacific pectinid species have hitherto been little studied at the population level, the above-mentioned favorable conditions and accumulation of knowledge should be put to full use in evolutionary studies of this family.

CHAPTER 2

Material and Method of Study

In order to detect geographic and chronological change of morphometric characters within an evolutionary lineage and to clarify morphological difference between lineages, many large random samples from various areas and horizons are indispensable. Although many fossil and Recent specimens of *Cryptopecten* are preserved for comparative studies in several museums, institutions and private collections in Japan, in many cases they are unsuitable for the present study because of meagre sample size, possibility of artificially biased sample preparation, or poorly documented locality and horizon. For the past ten years, therefore, I have endeavored to collect and observe as many sets as possible of large samples which faithfully represent local populations.

With the cooperation of a large number of scientists and amateurs I was able to examine 40 fossil samples and 30 Recent samples (including 115 subsamples) of *Cryptopecten vesiculosus*, as documented in the annexed list (p. 122). The total number of individuals comes to about 20,000. These samples appear to cover the main geographic and stratigraphic distributions of this species. The 17 samples treated in my 1973 paper were studied again and reanalyzed, because the biometric method of study and standardization of characters employed in the present study differ slightly from those used in the earlier one.

More than 270 living individuals of *C. vesiculosus* with soft parts were available for this study (Table 1). Most of them were obtained in six different seasons from sandy bottom near the edge of the continental shelf of eastern Sagami Bay. Though many individuals are immature (probably yearling), this sample, *Jg* (*1–26*), gives useful information about reproductive biology and population structure. The strictly sympatric relation of the two discrete phenotypes is also ascertained in this sample. This sampling locality was very convenient because of its proximity to the Misaki Marine Biological Laboratory of the University of Tokyo.

Another promising sea area for the investigation of living populations of *C. vesiculosus* is located near the mouth of Ise Bay, especially the sandy bottom about 10 km east of Cape Anori near Toba City. Though I have not yet succeeded in collecting living individuals, more than 300 fresh and conjoined valves (Sample *Is*) were collected from the wastes of a commercial fishery in this sea area by Mr. S. Hayashi. These are mostly mature and have proved very useful for the observation of general shell morphology (Plate 3, Fig. 5).

Several dated living specimens from Zenisu and two other sea banks west of Izushichitô Island (Samples *Zs*, *Hs* and *Ts*) provided by Dr. T. Okutani, and specimens from the southeastern part of Suruga Bay (Sample *Hk*) collected by Dr. H. Kitazato, are also important for the observation of shell growth because they were collected in several different seasons.

4

Table 1. Examined dated living specimens (mostly preserved with soft part) of *Cryptopecten vesiculosus*.

Sea area	Sample	Date	Depth	N	N_R	N_Q	Gonad
Uraga Strait	*Kz*	Oct. 4, 1969	60 m	2	1	1	unknown
E. Sagami Bay	*Jg (1)*	Oct. 4, 1979	80 m	10	8	2	undeveloped
ditto	*Jg (2)*	ditto	80–90 m	21	10	11	ditto
ditto	*Jg (3)*	Oct. 5, 1979	85–90 m	11	8	3	ditto
ditto	*Jg (4)*	ditto	85–90 m	18	10	8	ditto
ditto	*Jg (5)*	ditto	85–90 m	5	3	2	ditto
ditto	*Jg (6)*	ditto	80 m	4	1	3	ditto
ditto	*Jg (7)*	May 25, 1978	85 m	29	11	18	partly developed
ditto	*Jg (8)*	ditto	80 m	29	12	17	ditto
ditto	*Jg (9)*	ditto	80 m	33	16	17	ditto
ditto	*Jg (10)*	Mar. 11, 1982	85 m	2	0	2	undeveloped
ditto	*Jg (11)*	ditto	85 m	1	0	1	ditto
ditto	*Jg (12)*	ditto	82 m	2	2	0	ditto
ditto	*Jg (13)*	ditto	80 m	1	1	0	ditto
ditto	*Jg (14)*	ditto	80 m	1	1	0	ditto
ditto	*Jg (15)*	Mar. 12, 1982	85 m	1	0	1	ditto
ditto	*Jg (16)*	July 27, 1982	90 m	4	2	2	undeveloped (juv.)
ditto	*Jg (17)*	ditto	90 m	6	2	4	well developed
ditto	*Jg (18)*	ditto	90 m	5	5	0	ditto
ditto	*Jg (19)*	ditto	90 m	11	4	7	ditto
ditto	*Jg (20)*	ditto	90 m	19	8	11	ditto
ditto	*Jg (21)*	July 29, 1982	88 m	13	7	6	ditto
ditto	*Jg (22)*	Nov. 5, 1982	90 m	5	3	2	undeveloped (juv.)
ditto	*Jg (23)*	ditto	90 m	8	1	7	ditto
ditto	*Jg (24)*	Aug. 3, 1983	82–84 m	10	4	6	well developed
ditto	*Jg (25)*	ditto	84–85 m	10	6	4	ditto
ditto	*Jg (26)*	ditto	84 m	6	2	4	ditto
E. Sagami Bay	*Ma*	Mar. 12, 1982	105 m	1	1	0	undeveloped
W. Sagami Bay	*Aj (1)*	Dec. 12, 1962	90 m	1	1	0	unknown
ditto	*Aj (2)*	July 7, 1966	105 m	2	0	2	ditto
off Oshima	*Os*	Oct. 26, 1973	109–116 m	2	2	0	undeveloped
Zenisu Bank	*Zs (1)*	Oct. 4, 1965	113 m	1	0	1	unknown
ditto	*Zs (3)*	July 10, 1969	110 m	2	2	0	ditto
ditto	*Zs (4)*	ditto	90 m	4	2	2	ditto
ditto	*Zs (8)*	Oct. 3, 1971	106 m	2	1	1	ditto
Hyôtanse Bank	*Hs (2)*	July 12, 1967	120–125 m	1	0	1	unknown
Takase Bank	*Ts (2)*	Dec. 6, 1968	100 m	1	0	1	unknown
ditto	*Ts (3)*	July 11, 1969	120–200 m	1	0	1	ditto
Suruga Bay	*Hk*	Apr. 15, 1981	110 m	4	1	3	undeveloped
off Shirahama	*Sr*	June 15, 1983	140–160 m	4	0	4	undeveloped
Tanabe Bay	*Mb*	Mar. 21, 1974	unknown	1	0	1	undeveloped

N: Total number of individuals; N_R: Number of individuals belonging to Phenotype R; N_Q: Number of individuals belonging to Phenotype Q.

6

The fossil samples of *C. vesiculosus* are considerably biased in time and space. About half of them are from the Upper Pliocene and Pleistocene on Boso Peninsula, where at least seven horizons can be discriminated. In other areas, however, occurrence of large fossil samples is rather sporadic. Great improvements in the correlation and chronology of the Neogene and Quaternary deposits in Japan have been made over the past ten years through studies of widely distributed tephras, various radioactive isotopes, fission-track dating and paleo-magnetism in addition to biostratigraphic zones and microfossil datum planes. Though there is much room for further improvement, the absolute ages of many fossil beds yielding *Cryptopecten* samples have become much clearer. The absolute ages adopted in the present paper are mainly based on those given by Tsuchi (1981 ed.) for the Neogene and Lower Pleistocene and by Machida et al. (1971, 1974) and Sugihara et al. (1978) for the Middle-Upper Pleistocene of Boso Peninsula.

With these samples I attempt to restore the evolutionary pattern of *C. vesiculosus*. Restoration of phyletic evolution, I believe, should be achieved through three steps: examination of individual variation within each local population and its causal evaluation, recognition of geographic variation, the spatial integration, so to speak, of intrapopulational variation, and chronological integration of geographic variation. In the case of *C. vesiculosus*, although geographic variation in the geological past is still difficult to recognize, morphological difference outside the maximum range of geographic variation in Recent samples may be attributable to evolutionary change.

As to other Indo-Pacific extant species of *Cryptopecten*, only a small number of new samples were obtainable. As shown in the annexed list of examined samples, some specimens of *Cryptopecten bullatus* and *Cryptopecten nux* from south Japan were available for this study, but they were generally too small in sample size to permit detection of any biometrically significant morphological change with time. Much less is known about the geographic distribution and evolutionary change of the two extinct species, *Cryptopecten yanagawaensis* from the Middle Miocene of central and north Japan and *Cryptopecten spinosus* sp. nov. from the Upper Pleistocene of south Japan, because they are now represented only by samples of a few and a single fossil population, respectively.

Recently, I had the opportunity to examine the large collections of Recent pectinids in the National Museum of Natural History, Washington D. C. and the American Museum of Natural History, New York, through the courtesy of the curators of these museums. These collections include a large number of Recent samples of *C. bullatus* (commonly labelled *C. alli*) and *C. nux* (sometimes labelled *C. bernardi* or *C. corymbiatus*) from extensive areas of the Indo-Pacific as well as *Cryptopecten phrygium* from the western Atlantic. The samples collected by R/V Albatross of the U. S. Bureau of Fisheries in Southeast Asia, which are preserved in Washington, D. C., are well documented and were especially useful for this study. The geographic distribution and systematic description of these uncommon species are mainly based on my research on these collections. Nomenclatorial revision and synonymic assignment of some early described species of this genus was possible after my examination of Dr. T. R. Waller's unpublished data and photographs of the type specimens of these species in several European institutions.

In the present study all the known species of *Cryptopecten*, both fossil and living, were studied in order to understand the evolutionary history of this genus. Understanding of their interspecific relation and the process of speciation are, of course, the central aim, although owing to incomplete fossil records the method of study is often confined to conventional comparative morphology.

The following abbreviations are used for the indication of institutions where the specimens under discussion are preserved:

BM (NH):   Department of Zoology, British Museum (Natural History), London.

USNM: Department of Invertebrate Zoology (Molluscan Division), National Museum of Natural History, Smithsonian Institution, Washington, D. C.

AMNH: Department of Invertebrate Zoology, American Museum of Natural History, New York, N. Y.

IGPS: Institute of Geology and Palaeontology, Faculty of Science, Tohoku University, Sendai.

UMUT: Department of Historical Geology and Palaeontology, University Museum, University of Tokyo, Tokyo.

NSMT: Department of Zoology, National Science Museum, Tokyo.

CHAPTER 3

Geographic Distribution and Habitat
of Living *Cryptopecten*

Cryptopecten is represented by comparatively small and sometimes inconspicuous species, whose distribution and ecology have yet to be satisfactorily studied. It has been found mainly on sublittoral and upper bathyal sandy bottoms by scientists and fishermen using dredging and trawling techniques. It is primarily an Indo-Pacific genus, though there is an uncommon but authentic species in the Mexican Gulf region. Aside from *Propeamussium, Palliolum* and some other genera with thin and translucent shells, *Cryptopecten* is generally regarded as a pectinacean genus adapted to most deep-water environments.

In Japanese waters the occurrence of living *Cryptopecten* is restricted to sea areas under the influence of the Kuroshio warm current and its branch, the Tsushima warm current; there is no record of any representative species in Oyashio cold current areas or in semi-closed embayments (Fig. 1).

Much of our present knowledge about the distribution of outer sublittoral and upper bathyal benthic molluscs around the Japanese Islands derives from the "survey of the continental shelf bordering Japan", which was carried out during the years 1922–1930 largely by the R/V Sôyô-maru I. A part of the dredged bivalve shells were studied by Habe (1958), and all the organisms, which had hitherto been identified and recorded, were compiled in a comprehensive list by Horikoshi et al. (1982) with information about precise location, depth and bottom character. There were 658 Sôyô-maru I stations over an area covering the greater part of the continental shelf around the main part of the Japanese Islands (except for Hokkaido and Ryukyu Islands); "*Cryptopecten tissotii*" [=*C. bullatus*] and *C. vesiculosus* were recorded at 5 and 87 stations, respectively (Fig. 1). Although I have observed only a part of this enormous collection, the record has been useful in establishing the horizontal and bathymetric distribution of these species.

Dall, Bartsch and Rehder (1938) originally recorded *Cryptopecten alli* [=*C. bullatus*] from three stations of the U. S. Bureau of Fisheries' R/V Albatross around the Hawaiian Islands. The holotype of *C. alli* (USNM 173194), probably a solitary living specimen at that time, was collected from the coarse-grained sand and rocky bottom at 238–252 fathoms off the south coast of Oahu Island. Several lots of dead shells subsequently collected from the seas around Hawaii and the Philippines, deposited mainly in the National Museum of Natural History, Washington, D. C. and the Bernice P. Bishop Museum, Honolulu, were also from upper bathyal bottoms of similar or somewhat lesser depth. Dautzenberg and Bavay (1912) originally described a conjoined (probably living) specimen of *Cryptopecten bullatus* from Siboga Station 105 (275 m) off Sulu Archipelago, Philippines. *C. bullatus* seems to be distributed mainly in the central and northwest Pacific, but there is an aberrant occurrence of some dead shells (USNM 773983)

Table 2.　Examined dated living specimens of *Cryptopecten bullatus*.

Sea area	Sample	Date	Depth	N	Gonad
Hyôtanse Bank	*Hs* (*3*) (*B*)	Dec. 6, 1968	170 m	2	unknown
ditto	*Hs* (*9*) (*B*)	July 7, 1969	200 m	1	ditto
off Shirahama	*Sr* (*1*) (*B*)	June 15, 1983	140 m	1	undeveloped (juv.)
ditto	*Sr* (*2*) (*B*)	ditto	140–160 m	7	partly developed
ditto	*Sr* (*3*) (*B*)	ditto	180 m	3	partly developed

collected from the top of Nasca Ridge (228 m) about 1,200 km west of north Chile. Judging from the geographical position and somewhat specialized morphology, this sample may represent a local population semi-isolated in the region of Ekman's "East Pacific barrier" (Figs. 2 and 26).

C. bullatus is actually distributed in the lower sublittoral and upper bathyal waters off the Pacific and East China Sea coasts of southwest Japan, for it is evidently conspecific with what Japanese malacologists have traditionally called "*Cryptopecten tissotii* (Bernardi)".* As shown in Table 2, several young specimens with well-developed reproductive glands were collected from sandy bottom (140–180 m) off the coast of Shirahama, Kii Peninsula (sample *Sr* (*1–3*) (*B*)). According to Habe (1958) and Horikoshi et al. (1982), the present species was recorded from two stations (132 m and 183 m) off the eastern coast of Kii Peninsula and three stations (155 m, 219 m and 549 m) off the western coast of Kyushu. The examined material includes several specimens from Tsushima Strait (Sample *Ta* (*B*)), but, so far as I am aware, *C. bullatus* does not seem to live in the Japan Sea.

As was described by Okutani (1972), *C. bullatus* occurs commonly on three submarine banks, called Zenisu, Hyôtanse and Takase, to the west of the Izu-shichitô Islands. According to him, living individuals of this species were collected by the R/V Sôyô II at six stations (Samples *Hs* (*B*), etc.) ranging from 130 to 180–220 (or 160–260) meters in depth. The following living molluscs were said to be accompanied by *C. bullatus* in one or more dredge sample: *Gaza sericate* Kira, *Galeoastrea guttata* (Adams), *Serpulorbis xenophorus* Habe, *Sassia semitorta* (Kuroda and Habe), *Tritonoranella ranelloides* (Reeve), *Tucetona shinkuroensis* Hatai, Niino and Kotaka, *Nipponolimopsis decussata* (Adams), *Modiolus margaritaceus* (Nomura and Hatai), *Chlamys princessae* Kuroda and Habe, *Polynemamussium intuscostatum* (Yokoyama), *Lima fujitai* Oyama, *Lima profunda* Masahito, Kuroda and Habe, *Limatula* (*Stabilima*) *japonica* (Adams), *Meiocardia tetragona* (Adams and Reeve), *Chama* (*Amphicama*) *argentata* Kuroda and Habe.

On the same submarine banks *Cryptopecten vesiculosus* are common as well. According to Okutani (1972), living individuals of *C. vesiculosus* were collected at twelve stations (Samples *Zs* (*1, 3, 4, 8*), *Hs* (*2*), *Ts* (*2, 3, 5*)) ranging from 84 to 200 meters (see Table 1, Text-fig. 3). Dead shells of *C. vesiculosus* and *C. bullatus* were often found together in dredge samples, but living individuals of the two species were not found in association except for one sample (Station D52 by Okutani, 1972). The bathymetric ranges of the two

---

*　Though these Japanese specimens were recently referred to *Cryptopecten alli* Dall, Bartsch and Rehder, 1938 (Poutiers, 1981; Hayami, 1982), *C. alli* is evidently a junior synonym of *Pecten* (*Chlamys*) *bullatus* Dautzenberg and Bavay, 1912, as fully discussed in the systematic description (p. 96).

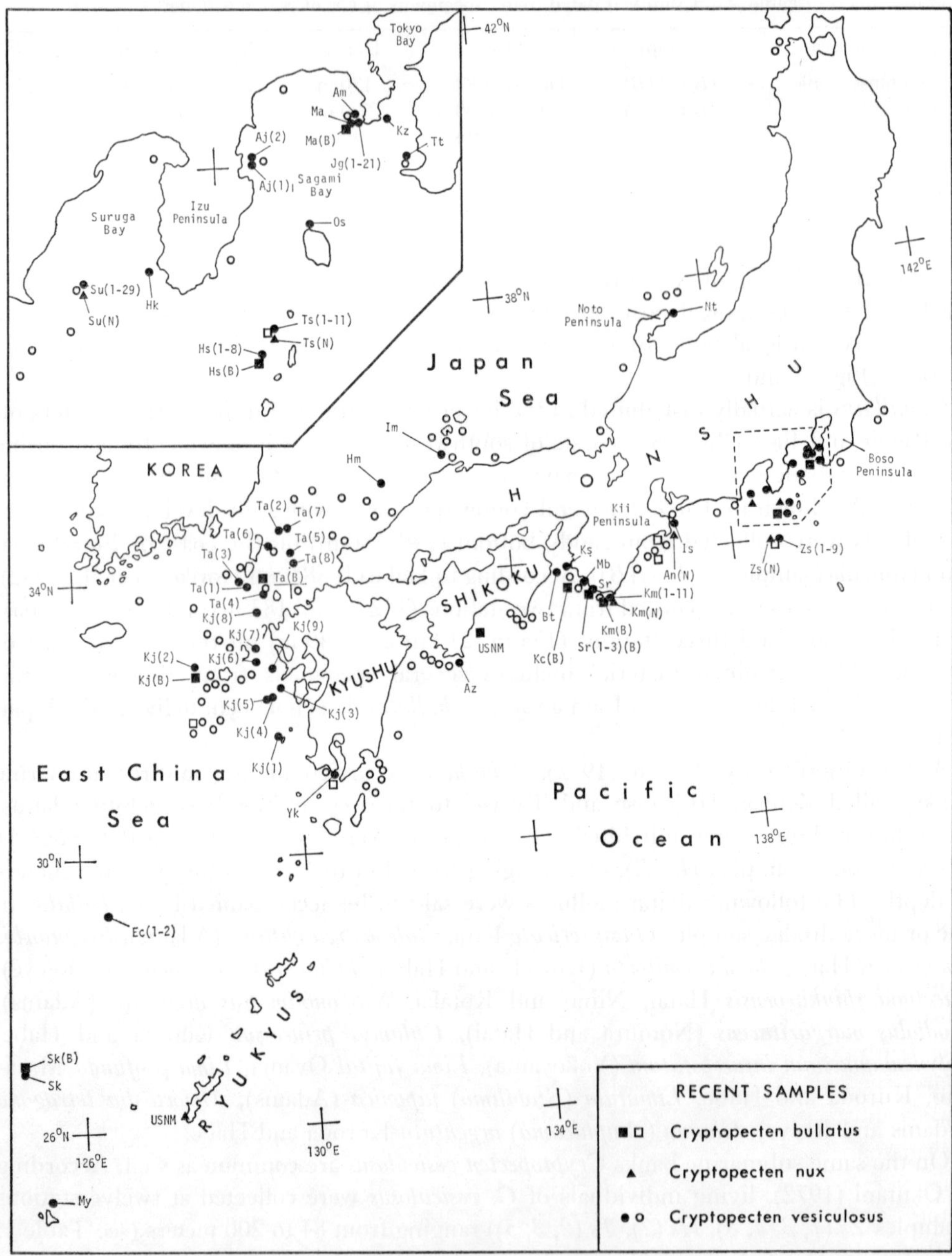

Fig. 1. Map showing the localities of Recent samples of *Cryptopecten* around the southwestern part of the Japanese Islands. *Aj*: off Ajiro, *Am*: Amadaiba, *An*: off Anori, *Az*: off Cape Ashizuri, *Bt*: off Benten Islet, *Ec*: west of Tokara Islands, *Hk*: off Cape Hakachi, *Hm*: off Hamada, *Hs*: Hyôtanse Bank, *Im*: off Izumo Peninsula, *Is*: off Anori, *Jg*: off Jôgashima Islet, *Kj*: west of Kyushu, *Km*: off Kushimoto, *Ks*: Kii Strait, *Kz*: off Cape Kenzaki, *Ma*: off Misaki, *My*: off Miyako Island, *Nt*: off Suzu, *Os*: off Okada, *Sk*: north of Senkaku Islets, *Sr*: off Shirahama, *Su*: Senoumi

species evidently overlap. Nevertheless, certain segregation between the populations of the two species is suggested. Taking bathymetric data in other sea areas around Japan into consideration, *C. vesiculosus* seems in general to prefer somewhat shallower bottom than *C. bullatus*. Okutani (1972) recorded the following living molluscs together with *C. vesiculosus* in one or more dredge samples: *Emarginula foveolata fujitai* Habe, *Gaza sericata* Kira, *Astralium okamotoi* Kuroda and Habe, *Galeoastraea girgylla* (Reeve), *Galeoastraea guttata* (Adams), *Tritonoranella ranelloides* (Reeve), *Biplex hirasei* Sowerby, *Chicoreus saltatrix* Kuroda, *Phos hirasei* Sowerby, *Tucetona shinkuroensis* Hatai, Niino and Kotaka, *Samacar pacifica* (Nomura and Zinbo), *Chlamys princessae* Kuroda and Habe, *Polynemamussium intuscostatum* (Yokoyama), *Lima fujitai* Oyama, *Meiocardia tetragona* (Adams and Reeve), *Chama* (*Amphicama*) *argentata* Kuroda and Habe, *Vasticardium arenicola* (Reeve), *Frigidocardium eos* (Kuroda), *Venus* (*Ventricolaria*) *toreuma* Gould.

The substrata of these submarine banks are, as described by Niino (1935, 1952), mainly gravel or coarse-grained sand with shell and coral fragments. Most of the above-listed molluscs seem to be epifaunal species inhabiting coarse-grained substrates of this sort.

Cryptopecten vesiculosus is a well-known and common species in southwest Japan. It is distributed not only off the Pacific coast of southwest Japan (35°N and south) and in the East China Sea but also off the Japan Sea coast of Honshu (41°N and south) through Tsushima Strait (Kuroda and Habe, 1952) (Fig. 1). According to the records of the R/V Sôyô I (Habe, 1958; Horikoshi et al., 1982), *C. vesiculosus* distributes predominantly on the continental shelf and slope around the southwestern half of the main part of the Japanese Islands (Honshu, Shikoku and Kyushu) from Boso Peninsula to Noto Peninsula. In the northern part of Honshu, however, this species was recorded only at two stations in the western part of Tsugaru Strait. The bathymetric range is said to be 51–582 m off the Pacific coast, 73–340 m off the East China Sea coast, and 73–406 m off the Japan Sea coast. The locations noted in these records are reliable but based mainly on the occurrence of dead shells including immature individuals and fragments, and the habitats of adult individuals joining reproductive activity may be restricted to somewhat narrower geographic and bathymetric ranges. Off the western coast of Miura Peninsula, for example, immature shells and, sometimes, living infant individuals can be collected from depths as shallow as 40 meters. On one occasion, near Cape Nagasaki-hana in Kagoshima Prefecture I obtained a few immature valves washed ashore (Sample *Yk*). So far as I am aware, the occurrence of living adult individuals is almost entirely restricted to the lower sublittoral zone between 60 and 200 meters.

Dautzenberg and Bavay (1912), on the Siboga Expedition, recorded *C. vesiculosus* from five stations in Indonesia. According to Dr. Dijkstra's personal communication (June 22, 1983) supported by photographs of the specimens in question, however, all of them undoubtedly belong to *C. nux*. The southern limit of main distribution of *C. vesiculosus* probably lies between Ryukyus and Formosa. The examined samples, therefore, seem to cover roughly the known distribution of this species.

Bank, *Ta*: around Tsushima Islands, *Ts*: Takase Bank, *Tt*: Tateyama Bay, *Yk*: off Cape Nagasakihana, *Zs*: Zenisu Bank. USNM: Localities of samples in the National Museum of Natural History, Washington, D. C. open mark: other known occurrence.

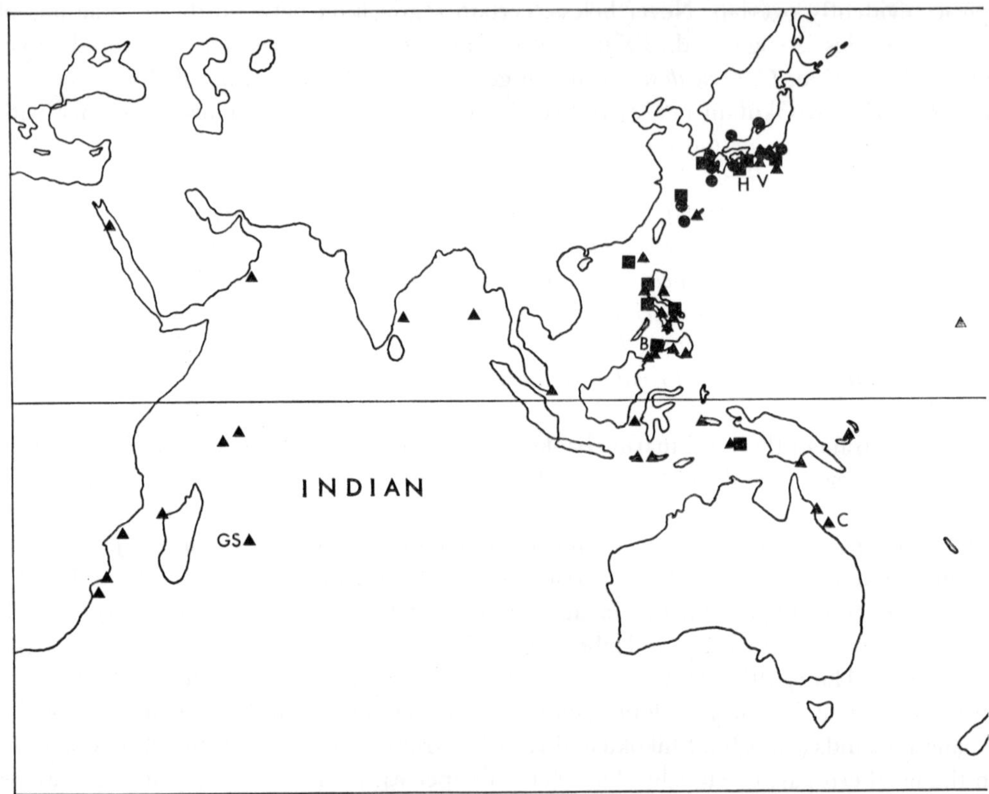

Fig. 2. Map showing the distribution of four extant species of *Cryptopecten*. Plots outside Japan are mainly based on the collections in the National Museum of Natural History, the American Museum of Natural History and the Amsterdam Zoological Museum. *A*: type locality of *Cryptopecten alli* [Oahu], *B*: type locality of *Pecten (Chlamys) bullatus* [Sulu Archipelago], *C*: type locality of *Chlamys corymbiata* [Queens-

In the eastern part of Sagami Bay near the southwestern coast of Miura Peninsula, *C. vesiculosus* is one of the commonest bivalves on sandy substrates (Biological Laboratory, Imperial Household, 1971; etc.). After a number of dredge trials on board the R/V Rinkai of the Misaki Marine Biological Laboratory, the University of Tokyo, I succeeded in determining the habitat of adult individuals and collected as many as 260 living specimens, almost all from the same station, about 2.5 km west of the western end of Jôgashima Islet (Table 1). In this sea area the periphery of continental shelf is only 100 meters deep, probably owing to intermittent upheaval. *C. vesiculosus* dwells near the shelf periphery where rich nutrition seems to be supplied by upwelling water. The following living molluscs were found together in one or more dredge samples: *Emarginula fragilis* Yokoyama, *Enida japonica* Adams, *Turcica coreensis* Pease, *Tristichotrochus haliarchus* (Melvill), *Tristichotrochus aculeatus* (Sowerby), *Galeoastraea modesta* (Reeve), *Serpulorbis xenophorus* Habe, *Balcis martinii* (Adams), *Xenophora japonica* Kuroda and Habe, *Polinices sagamiensis* Pilsbry, *Tritonoranella ranelloides* (Reeve) (only the phenotype with large nodes), *Reticutriton tenuiliratus* (Lischke), *Charonia sauliae macilenta* Kuroda and Habe,

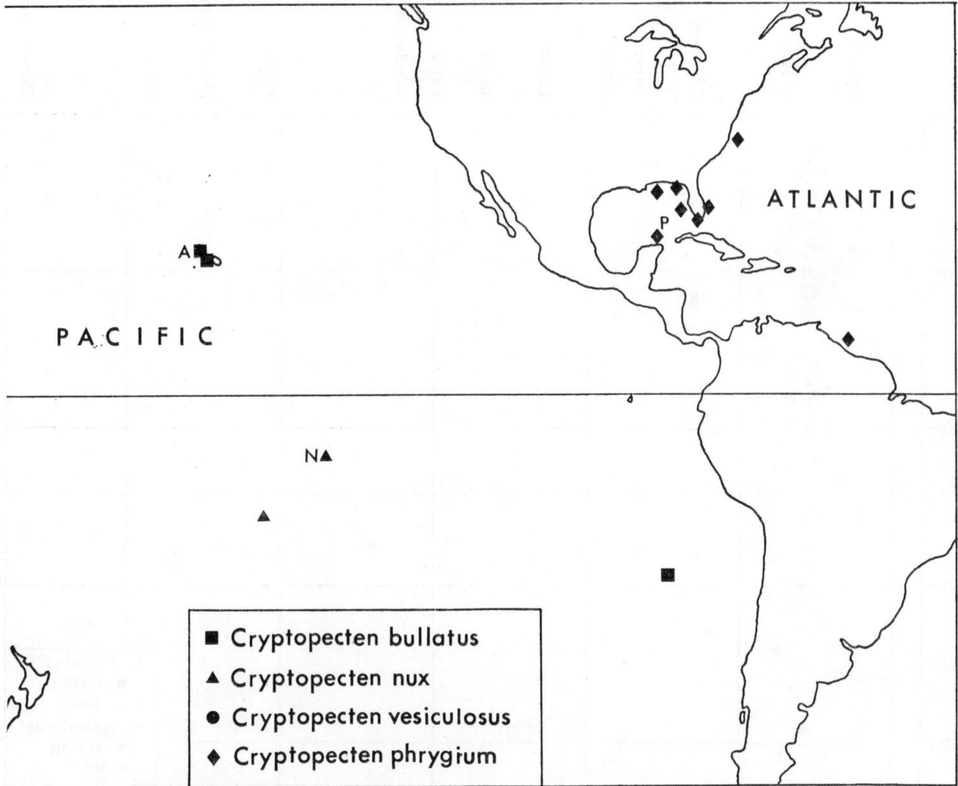

land], *G*: type locality of *Pecten guendolenae* [Mauritius], *H*: type locality of *Pecten hastingsii* [Japan], *N*: type locality of *Pecten nux* [Marquesas], *P*: type locality of *Pecten phrygium* [Mexican Gulf], *S*: type locality of *Chlamys smithi* [Mauritius], *V*: type locality of *Pecten vesiculosus* [Japan].

Ocenebrellus aduncus (Sowerby), *Pteropurpura vespertilis* Kuroda, *Fusinus perplexus perplexus* (Adams), *Granulifusus* sp., *Phos hirasei* Sowerby, *Endemoconus sieboldii* (Reeve), *Glycymeris rotunda* (Dunker), *Crenulilimopsis oblonga* (Adams), *Chlamys lemniscata* (Reeve), *Polynemamussium intuscostatum* (Yokoyama), *Lima fujitai* Oyama, *Neopycnodonte musashiana* (Yokoyama), *Megacardia ferruginosa* (Adams and Reeve), *Meiocardia tetragona* (Adams and Reeve), *Nemocardium* (*Keenaea*) *samarangae* (Makiyama), *Frigidocardium eos* (Kuroda), *Pitar* (*Pitarina*) *affine* (Gmelin), *Venus* (*Ventricolaria*) *foveolata* (Sowerby), *Pandora* (*Pandorella*) *wardiana* (Adams), *Cuspidaria chinensis* (Griffith and Pidgeon).

At this station *C. vesiculosus* is the most dominant molluscan species, but a reddish terebratellid brachiopod, *Laqueus rubellus* (Sowerby) is found in greater abundance. *Crenulilimopsis oblonga, Lima fujitai, Megacardia ferruginosa, Turcica coreensis* and *Tritonoranella ranelloides* are also quite abundant. The substratum at this station is also coarse-grained sand containing a considerable amount of fine gravel and shell fragments. Living individuals of *C. vesiculosus* were mostly detached from such hard objects when dredged, but in some cases they were clearly attached to gravel and shell fragments by

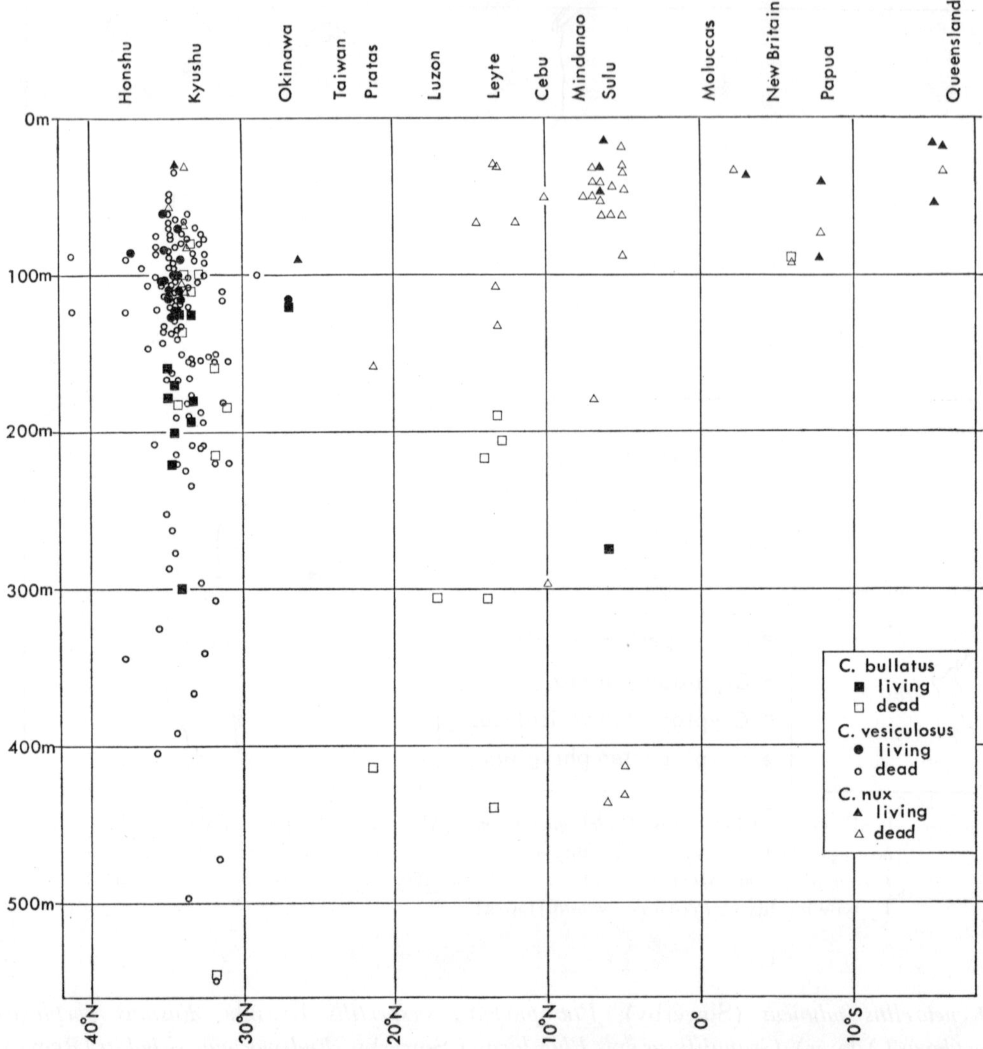

Fig. 3. Bathymetric distribution of *Cryptopecten* in the western Pacific. Latitudinal
and bathymetric data are mainly derived from the results of Soyo-maru, Siboga and
Albatross expeditions in addition to the present samples.

means of delicate byssus.

No one has successfully observed the actual life of *C. vesiculosus* in its habitat. Judging
from their behavior in aquaria, striking contrast of pigmentation between two valves and
more abundant attaching organisms on the left valve, however, it is almost certain that
living individuals of this species, like many other byssate pectinids, lie on the sea floor with
the left valve facing up.

Molluscs and other organisms associated with *C. vesiculosus* vary considerably with
location, and there is thus no definite assemblage of species. In spite of what many have
written on molluscan synecology, I feel that a "community", especially in open sea

environments, is generally nothing but an accidental coexistence of different species, except for special cases such as parasitism and commensalism. Sometimes ahermatypic corals, barnacles, bryozoans and calcareous algae grow on the surface of living shells. Coexisting species of *Chlamys* are often encrusted with sponges, but I have never found this phenomenon in *C. vesiculosus*.

Cryptopecten nux seems to be rare in the present-day seas around the Japanese Islands, and there is no record of this species in the northern part of the East China Sea and the Japan Sea. Because of the paucity of well-documented records, its geographic and bathymetric distributions in the Japanese waters remain obscure. To date living specimens from southern Japan have been collected largely from upper sublittoral sandy bottom at depths less than 50 meters, though dead shells have also been found in some samples from the outer sublittoral zone (e. g., samples *Zs* (*N*), *Km* (*N*)), together with *C. vesiculosus*.

C. nux is extensively distributed in the tropical Indo-Pacific; there are about 140 samples of this species in the National Museum of Natural History, Washington D. C., from various areas of Polynesia, Melanesia, Micronesia, Philippines, Borneo, South China Sea, Okinawa, Singapore, Queensland, Andaman, Seychelles, Mauritius, Madagascar, Mozambique, eastern South Africa, and the Red Sea. They were collected from various depths of sublittoral and bathyal zones, but the bathymetric range of living specimens seems to be confined to 30–100 m (Figs. 2 and 3).

Cryptopecten phrygium is the solitary Atlantic representative of this genus. As was noted in the original description and is represented by the collections of the National Museum of Natural History and the American Museum of Natural History, this species lives at about 200 meters in the Mexican Gulf and adjacent seas. Most specimens in these museums came from the sea off Florida, but this species is known to inhabit a wide area of the western Atlantic and has been found off Georgetown in Guiana, off the Yucatan Peninsula, and off the eastern coast of the United States from Florida to Cape Cod (Fig. 2).

CHAPTER 4

Stratigraphic Distribution and Mode of Occurrence
of Fossil *Cryptopecten*

Cryptopecten is an extensively distributed genus, but the occurrences of fossils so far recorded seem to be restricted almost entirely to the Neogene and Quaternary of Japan except for some fossils of *C. nux* in a few regions of Southeast Asia and east Africa. Reflecting the influence of the warm paleocurrents, the occurrences of fossil *Cryptopecten* are concentrated in the Pacific coastal region of southwest Japan and the Ryukyu Islands except for a few cases on the Japan Sea coast (Fig. 4).

Fossil *Cryptopecten* occur dominantly in coarse-grained, sometimes conglomeratic, sandy sediments. Warm open-sea elements of molluscs commonly occur together. Conjoined valves are quite rare, and at some localities many valves suffered strong abrasion before burial. I have not seen any strictly autochthonous occurrence of fossil *Cryptopecten*, but judging from the lithology of fossil beds and associated molluscan fossils, the habitat of *Cryptopecten* in geological ages after Middle Miocene does not seem to be much different from that of the present time.

Two Early Miocene (formerly regarded as Late Oligocene) species, *Pecten kyushuensis* Nagao, 1928, from the Waita Formation in Fukuoka Prefecture (north Kyushu) and *Aequipecten hataii* Kanno, 1958, from the Nenokami Formation in Saitama Prefecture (central Honshu), are possibly early representatives of *Cryptopecten*, but their phylogenetic relation to Middle Miocene and later species is by no means certain owing to the poorly preserved nature of the material.

Pecten (Aequipecten?) yanagawaensis Nomura and Zinbo, 1936, from the early Middle Miocene (15–16 Ma) Yanagawa Formation in Fukushima Prefecture (north Honshu) is an extinct species safely assignable to *Cryptopecten* because of the many essential characteristics it shares with *C. bullatus* and *C. vesiculosus*. *C. yanagawaensis* is also known from the Moniwa Formation in Miyagi Prefecture (Nomura, 1940; Masuda, 1958, 1962), the Shukunohora Formation in Gifu Prefecture (Itoigawa, Shibata and Nishimoto, 1974) and the Sunakozaka Formation in Ishikawa Prefecture (Ogasawara, 1976), all in north-central Honshu and early Middle Miocene in age. During early Middle Miocene times (approximately 15–16 Ma) a tropical or subtropical climate predominated, and a characteristic molluscan fauna of warm-current type spread over the Japanese Islands except for the northern part of Hokkaido (Chinzei, 1978). *C. yanagawaensis* is undoubtedly an element of the warm-water fauna of this age.

After middle Middle Miocene, tropical-subtropical faunas seem to have retreated from the main part of the Japanese Islands, and there are few fossil records of *Cryptopecten* in the Late Miocene and Early Pliocene sediments. Shikama (1973, pp. 190–191) recorded the occurrence of several specimens of *C. yanagawaensis* in the basal part of the Zushi Formation, which is probably Late Miocene in age. This occurrence may be important

16

for the present study, but the specimens were not described and unfortunately cannot be found in the collection he bequeathed to the Yokohama National University.

The earliest occurrence of *Cryptopecten vesiculosus* is found in the Middle Pliocene (approximately 3.5 Ma) calcareous sandstone of the Harada Formation of the Shirahama Group in Shizuoka Prefecture (central Honshu). This species seems to have flourished from Late Pliocene onwards, often constituting a dominant element in the molluscan faunas of Kuroshio type. There are numerous fossil localities in the Upper Pliocene and Pleistocene along the Pacific coast of southwest Japan, and a few small samples were also obtained from some Plio-Pleistocene deposits on the Japan Sea coast. In certain period of Pliocene this species also distributed on the Pacific coast of north Honshu beyond the northern limit of distribution at the present time.

In the Boso Peninsula in Chiba Prefecture (near Tokyo) thick marine deposits of Late Pliocene and Pleistocene are almost successively developed. The molluscan faunas seem to have changed in harmony with the oscillations of warm and cold currents, and there are at least seven horizons yielding *C. vesiculosus*; namely, Kurotaki Formation (Sample *Iy*), Umegase Formation (Sample *Tm 1, 2*), Ichijuku Formation (Sample *Ij 1, 2*), Sanuki Formation (Sample *Sn 1–3*), "Higashiyatsu Formation" (Sample *Hg*), Jizôdô Formation (Sample *Ny 1–4, Jz, Ic, Ab 1, 2*), Yabu Formation (Sample *Sm*), Kioroshi Formation (Sample *Tn*) in upward sequence. The sequence and stratigraphic position of samples are collectively shown in Fig. 5. Though such a good sequence of marine sediments is rather exceptional, Late Pliocene and Early Pleistocene samples of *C. vesiculosus* were obtained from the Miura Group (Sample *Nj*) in Kanagawa Prefecture (central Honshu), the Kakegawa Group (Sample *Kg 1, 2*) in Shizuoka Prefecture (central Honshu), the Miyazaki Group (Sample *Mz 1, 2*) in Miyazaki Prefecture (south Kyushu), the Shimajiri Group (Sample *Ik, Ob*) on Okinawa Island, and a few other areas.

Cryptopecten-bearing fossil beds of Late Pleistocene and younger age seem to be rare in southwest Japan, probably because outer sublittoral sediments have not been raised on land over such a short period. In the inner part of Kagoshima Bay and its surrounding volcanic region (south Kyushu), however, local uplift of an unusual rate must have occurred, and a striking Holocene fossil bed (ca. 2,720 years B. P.) with abundant valves of *C. vesiculosus* is exposed at Moeshima (=Niijima) Islet (Sample *Ms*).

The fossil record of *Cryptopecten nux* in Japan also dates back to the Middle Pliocene. It occurs, though rarely, in the Yonahama Formation of the Shimajiri Group (approximately 3.0 Ma) at Miyako Island, Okinawa Prefecture (Sample *Mk (N)*), and also in the Late Pliocene Shinzato Formation (Sample *Ik (N)*) at Ikei Islet and the Early Pleistocene Chinen Formation (Sample *Ob (N)*) in the southern part of Okinawa Island. At the second and third localities *C. nux* is found together with *C. vesiculosus*. *C. nux* seems in fact to be a long-lived species, the occurrence of some Early Miocene fossils having been confirmed in Tanzania (Eames and Cox, 1956).

Since Nomura and Zinbo (1934) described the rich molluscan fauna of the Ryukyu (=Riukiu) Limestone at the north of Kamikatetsu on Kikai Island, Kagoshima Prefecture, many investigators have visited this locality for various purposes of stratigraphy and paleontology. According to Sakanoue et al. (1967) and Omura (1983), this fossil bed is about 80,000 years old. Nomura and Zinbo reported *Pecten (Aequipecten) kikaiensis* together with *Pecten (Aequipecten) vesiculosus* and *Pecten (Aequipecten) nux*. As a result of

18

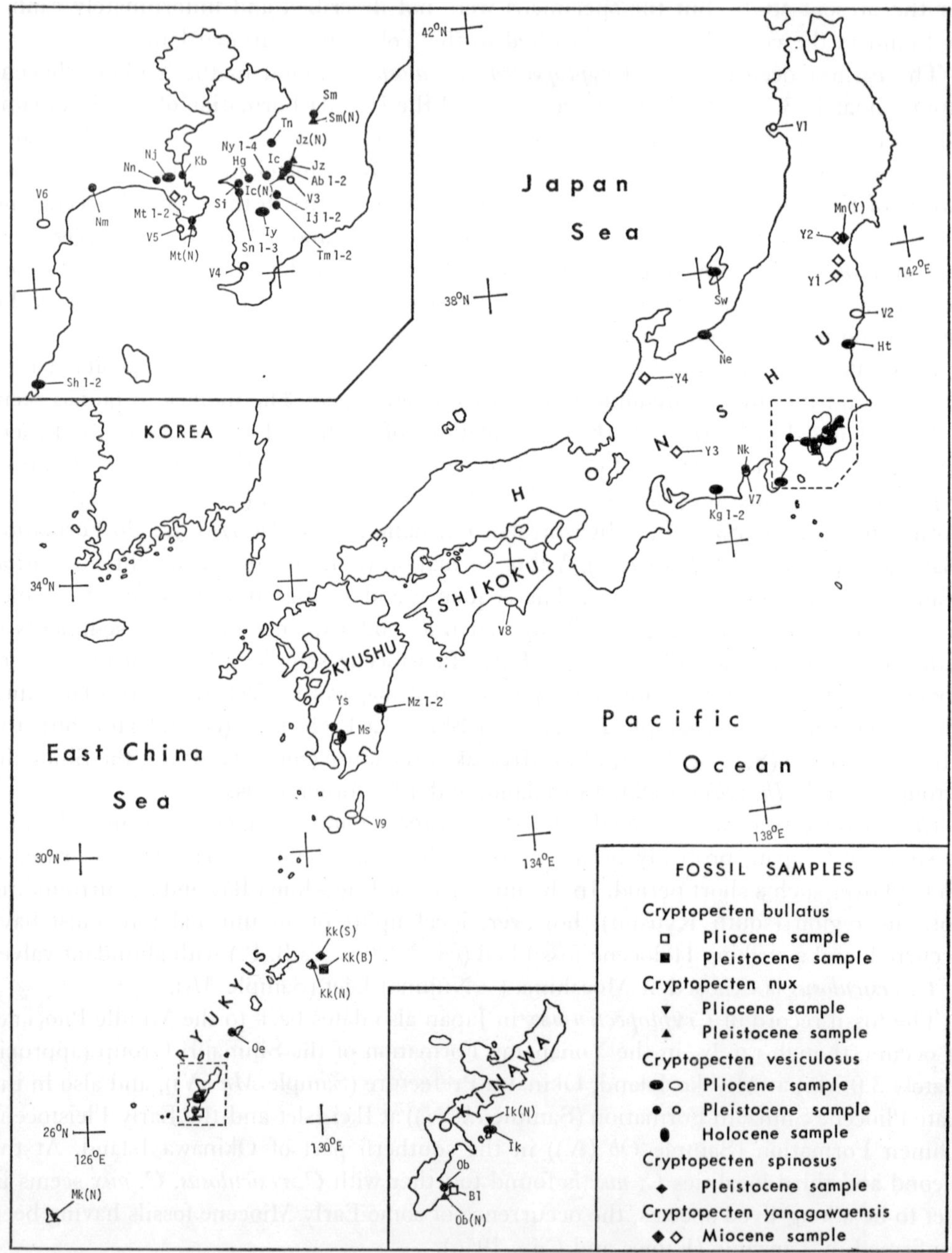

Fig. 4. Map showing the localities of studied fossil samples of *Cryptopecten* in the Japanese Islands. *Ab*: Jizôdô Formation at Atebi, *Hg*: Jizôdô Formation at Kori, *Ht*: Taga Group at Hitachi, *Ic*: Jizôdô Formation at Ichinosawa, *Ij*: Ichijiku Sand at Shibahara, *Ik*: Shinzato Formation at Ikei Islet, *Iy*: Kurotaki Formation at Ishiyama, *Jz*: Jizôdô Formation at Jizôdô, *Kg*: Hosoya Silt at Ugari, *Kk*: Wan Formation at Kikai Island, *Mn*: Moniwa Formation at Junishin, *Ms*: Moeshima Islet (subfossil), *Mt*: Miyata Formation at Tsukui, *Mz*: Miyazaki Group at Hagenoshita and Tori-

my survey on their original specimens in the Tohoku University and newly collected material, I have reached somewhat different taxonomic conclusions. The holotype of *P. (A.) kikaiensis* (IGPS coll. cat. no. 50357) is evidently conspecific with the numerous newly collected specimens from the same locality (Sample *Kk (N)*) which are safely assignable to *Cryptopecten nux*. At the same locality there are many specimens belonging to two other species of *Cryptopecten*. One is, as Nomura and Zinbo (1934) judged, certainly related to *C. vesiculosus* but clearly distinguishable from it by the fewer and more spinose radial ribs and weaker convexity of left valve, as described in this paper under the name of *Cryptopecten spinosus* sp. nov. (Sample *Kk (S)*). The other species, represented by weakly inflated and thin-shelled specimens (Sample *Kk (A)*), is identical with *Cryptopecten bullatus*, though all the specimens are regarded as immature. Incidentally, no fossils of *C. bullatus* have been recorded from any other locality in Japan (or probably, anywhere else in the world) except for a few comparable specimens from the Late Pliocene Shinzato Formation of Okinawa Island, described by Noda (1980) under the name of "*Aequipecten* sp."

Yokoyama (1922) described two strongly inflated valves of a small pectinid from "Shitô" [Late Pleistocene Yabu Formation at Ochishimoshinden of Semata, Chiba Prefecture] under the name of *Pecten tissoti* [should be *tissotii*]. In the revisory monographs of Yokoyama's illustrated specimens, Taki and Oyama (1954) and Oyama (1973) called it *Aequipecten (Cryptopecten) sematensis* Oyama. This form is, as described later, here treated as a subspecies of *Cryptopecten nux*.

---

yamahama, *Ne*: Byôbudani Formation at Naoetsu, *Nj*: Nojima Formation at Totsuka, *Nk*: Nekoya Formation at Furuyado, *Nm*: Ninomiya Formation at Mushikubo, *Nn*: Naganuma Formation at Tengakuin, *Ny*: Jizôdô Formation at Nishiyatsu, *Ob*: Chinen Formation at Shimo-oyakebaru, *Oe*: Shimoshiro Formation at Okinoerabu Island, *Sh*: Harada Formation at Shirahama, *Si*: Sunami Formation at Nishi-ôwada, *Sm*: Yabu Formation at Ochishimoshinden, *Sn*: Sanuki Formation at Sasage, *Sw*: Sawane Formation at Sawada, *Tm*: Umegase Formation at Tsujimori, *Tn*: Kioroshi Formation at Iitomi, *Ys*: Yoshida Formation at Unoki. Open marks indicate other confirmed fossil occurrences; *Al*: Shinzato Formation at Chinen, *V1*: Shibikawa Formation at Oga, *V2*: Taga Group at Futatsunuma, *V3*: Mandano Formation at Mandano, *V4*: "Kiwada Formation" at Nako, *V5*: Miyata Formation at Hatsuse, *V6*: Hayakawa Tuff at Sukumogawa, *V7*: Nekoya Formation at Nakahiramatsu, *V8*: Nobori Formation at Nobori, *V9*: Tajima Formation at Kajigata, *Y1*: Yanagawa Formation at Yanagawa, *Y2*: Moniwa Formation at Moniwa, *Y3*: Akeyo Formation at Mizunami, *Y4*: Sunakozaka Formation at Kanazawa.

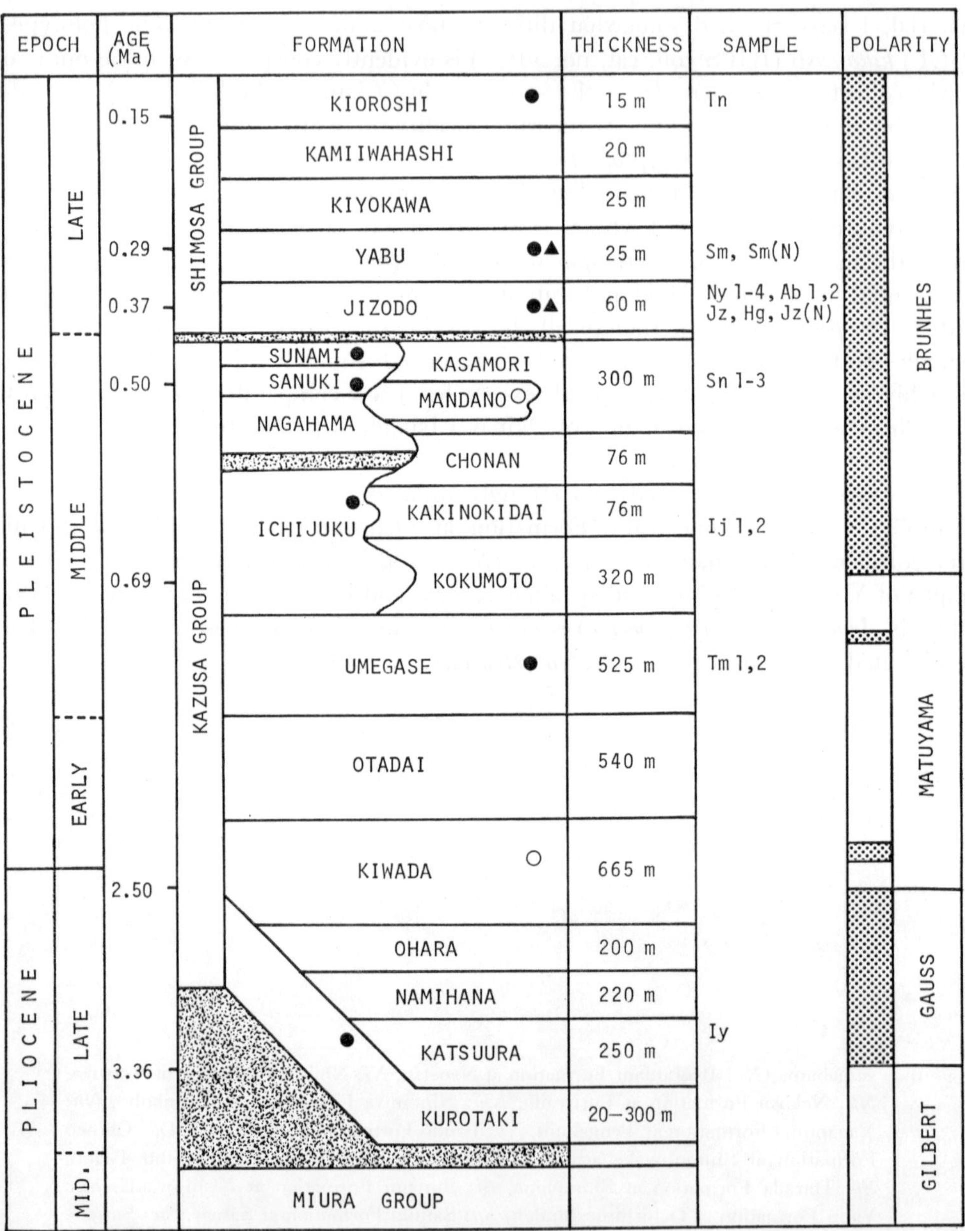

Fig. 5. Stratigraphic position of *Cryptopecten* samples in the Plio-Pleistocene sequence of Boso Peninsula near Tokyo. ●: *C. vesiculosus*, ○: *C. vesiculosus* (Horizon unknown in detail), ▲: *C. nux*. The stratigraphic division and thickness of strata are mainly adapted from Mitsunashi *et al.* (1979), and the absolute ages and magnetic polarity from various sources.

CHAPTER 5

Ontogenetical Development of Surface
Sculpture in *Cryptopecten*

The surface sculpture of *Cryptopecten* is unique and clearly different from that of other pectinid genera. Because the present evolutionary study depends to a great extent upon the variability of radial ornamentation, it will be necessary to clarify the detailed structure and ontogenetical development of surface morphology in each representative species.

The surface morphology of the following nine selected individuals was examined by use of a scanning electron microscope (Hitachi: S–700) with relatively low magnification.

Specimen 1 (UMUT RM16002a). Conjoined (living) valves of *Cryptopecten bullatus*, belonging to Sample *Hs (3) (B)*. 13.2 mm long, 12.4 mm high, 3.8 mm thick. [Plate 10, Fig. 3; Plate 11, Fig. 3]

Specimen 2 (UMUT RM16059a). Conjoined (living) valves of *Cryptopecten vesiculosus* [Phenotype *R*]*, belonging to Sample *Hy*. 10.6 mm long, 11.0 mm high, 4.4 mm thick. [Plate 10, Fig. 2; Plate 11, Fig. 2]

Specimen 3 (UMUT RM16059b). Conjoined (living) valves of the same species [Phenotype *Q*]*, belonging to the same sample. 11.6 mm long, 11.8 mm high, 5.0 mm thick. [Plate 10, Fig. 1; Plate 11, Fig. 1]

Specimen 4 (UMUT RM16061a). Conjoined (living) valves of the same species [Phenotype *R*], belonging to Sample *Jg (2)* (for the observation of transverse section and ventral margin of adult shell). 25.3 mm long, 25.6 mm high, 11.8 mm thick. [Plate 12, Fig. 6; Plate 13, Fig. 2]

Specimen 5 (UMUT RM16138a). Conjoined (living) valves of the same species [Phenotype *Q*], belonging to Sample *Is* (for the observation of transverse section and ventral margin of adult shell). 27.3 mm long, 25.4 mm high, 12.7 mm thick. [Plate 13, Fig. 1]

Specimen 6 (UMUT CM16015a). Right (fossil) valve of *Cryptopecten nux*, belonging to Sample *Kk (N)*. 12.2 mm long, 12.1 mm high, 3.5 mm thick. [Plate 12, Fig. 1]

Specimen 7 (UMUT CM16015b). Left (fossil) valve of the same species, belonging to the same sample. 12.0 mm long, 12.1 mm high, 2.7 mm thick. [Plate 12, Fig. 2]

Specimen 8 (UMUT CM16170a). Right (fossil) valve of *Cryptopecten spinosus* sp. nov., belonging to Sample *Kk (S)*. 12.5 mm long, 12.3 mm high, 3.0 mm thick. [Plate 12, Figs. 3, 5]

Specimen 9 (UMUT CM16170b). Left (fossil) valve of the same species, belonging to the same sample. 12.9 mm long, 13.2 mm high, 3.0 mm thick. [Plate 12, Fig. 4]

Because the umbonal area of full-grown shells is commonly abraded or concealed by encrusting organisms, suitable specimens for the observation of early stages are restricted to young individuals such as those of specimens 1–3, 6–9.

As the young shells of *C. bullatus*, *C. vesiculosus* (both phenotypes), *C. nux* and *C.*

---

* For definition of Phenotype *Q* and *R*, see the description and discussion on intrapopulational variation (pp. 52–60).

22

spinosus share many common surface characteristics, the ontogenetical development of surface sculpture is collectively described below.

Prodissoconch.—The prodissoconch is well demarcated on the umbonal surface in these young individuals (Specimens 1–3, 6–9). It is D-shaped in outline without wings, strongly inflated in both valves and as large as 140–200 microns in length. The surface is apparently smooth. *C. bullatus* seems to have a slightly larger prodissoconch than three other species. As is commonly the case in bivalves (e.g., Ockelmann, 1956; Waller, 1981), the prodissoconch is generally composed of two parts: a smooth or radially striated part near the origin of growth (prodissoconch I) and a commarginally striate part (prodissoconch II). The two parts are not clearly distinguishable in these examined specimens, but this may be due to subsequent abrasion. [See Plates 10 and 12]

First stage of dissoconch (shell height less than 1.5 mm).—The shell of this stage is strikingly inequivalve and inequilateral, left-convex and pteriform with indistinctly demarcated posterior wing. The outline reminds one of some aviculopectinids from the Upper Paleozoic and Lower Mesozoic. The byssal notch becomes gradually distinct, but a ctenolium is still undeveloped. The disk and wings of the right valve are smooth except for weak growth-lines. In contrast, the left valve is ornamented with numerous *Camptonectes*-like divaricate striae which are also distributed on both wings. In the later half of this stage radial ribs appear on the disk of the left valve, though their number and strength are still unstable owing to irregular insertion, convergence and disappearance. [See Plates 10 and 12]

Second stage of dissoconch (shell height between 1.5 mm and 4 mm).—In this stage the outline of shell changes from pteriform to pectiniform and from left-convex to equiconvex. The posterior wing becomes well delimited, and several denticles of the ctenolium are developed along the anterodorsal margin of disk. *Camptonectes*-like striae are persistent on the disk and wings of the left valve and appear also on the disk and posterior wing of the right valve. Radial ribs develop on the disk of the right as well as the left valve. They are sometimes flexuous but simple, provided with fine scales, neither bifurcated nor inserted, and always much narrower than the interspaces. Their number varies with the species (in the examined specimens, 22 in *C. bullatus*, 16 or 17 in *C. vesiculosus* and 22 in *C. nux*). A few much broader radial ribs appear on the wings of both valves, and on the left anterior wing they form a striking lattice sculpture with strong concentric lamellae. [See Plates 10 and 12]

Third stage of dissoconch (shell height more than 4 mm).—The shell is nearly acline and equiconvex in the earlier half of this stage but becomes right-convex and slightly prosocline in the later. Denticles of the ctenolium continue to develop throughout this stage. The *Camptonectes*-like striae disappear entirely by the beginning of this stage. Radial ribs are persistent except for one or two on the peripheral areas of the disk, relatively narrow at first but apparently much broadened later. The broadened radial ribs show a very unique structure. As clarified by SEM, each broadened rib consists of a central solid ridge and a pair of lateral hollow parts. The solid part is virtually the continuation of a radial rib of earlier stages, but the hollow parts occur only in the later half of this stage. The hollow parts are originally covered with numerous opposite or alternate imbricated scales, which are, however, apt to be ragged and exfoliated even in living shells. Consequently, in fossils and water-worn dead shells radial ribs often look as if they were tripartite or accompanied by a pair of subordinate riblets. Fine erect scales occur on the interspaces of radial ribs, and their periodicity are quite independent from the scales on the ribs. The mode of radial ornamentation in the later half of this stage varies considerably with the constituent species and is also different between the two phenotypes of *C. vesiculosus*, as separately shown in the systematic description (p. 107). [See Plate 11]

The pteriform and left-convex outline of early dissoconch seems to be ubiquitous in living pectinids. The *Camptonectes*-like divaricate striae are also widespread in the Pectinacea (Waller, 1972a), though they vary in prominence and persistence with the species. In the Pectinacea, according to Waller (1972b), there is a close correlation between surface sculpture and shell microstructure. Waller stated that various conspicuous sculptures (radial ribs, spines, ctenolium and *Camptonectes*-like striae) occur only on surfaces composed of foliated calcite, while a prismatic calcite layer, which occurs exclusively on the external surface of the right valve (throughout growth in *Propeamussium*, but only on early dissoconch in *Chlamys* and *Amusium*), generally forms a smooth surface. This is certainly true in the examined species of *Cryptopecten*, the first appearance of *Camptonectes*-like striae and radial ribs being much earlier in the left valve. The external surface of the first stage of dissoconch probably consists of prismatic calcite in the right valve and of foliated calcite in the left valve.

CHAPTER 6

Reproductive Season and Growth Rings
of *Cryptopecten*

Various kind of basic information about the reproductive strategy and life cycle in living populations is of prime importance in order to understand the evolution of fossil organisms. Since *Cryptopecten* is mainly composed of lower sublittoral and bathyal species, direct embryological observation is difficult. My knowledge on this matter is still restricted to the anatomical features and their relation to shell morphology in some dredged samples of *C. vesiculosus* and *C. bullatus*.

As shown in Table 1, numerous living individuals of *C. vesiculosus*, belonging to 26 subsamples, were collected from a sandy bottom—almost all at the same station—about 2.5 km west of the western end of Jôgashima Islet in the eastern part of Sagami Bay [sample *Jg*, 35°07.9′N, 139°35.1′E, 80–90 m].

They were obtained in the following six different seasons:

(1) October 4 and 5, 1979 [Subsample *Jg* (*1–6*)]
(2) May 25, 1978 [Subsample *Jg* (*7–9*)]
(3) March 11 and 12, 1982 [Subsample *Jg* (*10–15*)]
(4) July 27 and 29, 1982 [Subsample *Jg* (*16–21*)]
(5) November 5, 1982 [Subsample *Jg* (*22, 23*)]
(6) August 3, 1983 [Subsample *Jg* (*24–26*)]

Though the greater part of them are immature and probably yearling, all the large individuals collected in summer possess well developed gonad. Therefore, this station can definitely be regarded as being situated in an area where reproduction actually occurs.

The Pectinidae are sometimes said to have sexes represented in separate individuals (Drew, 1906) but to be more commonly hermaphroditic. Hermaphroditic reproductive glands are known in *Pecten* (*Pecten*) *maximus* and *Aequipecten opercularis* from England (Dakins, 1909; Rees *in* Cox, 1957; etc.) and also in two commercially important Japanese species, *Pecten* (*Notovola*) *albicans* and *Patinopecten* (*Mizuhopecten*) *yessoensis*. They have also been found in *C. vesiculosus* and *C. bullatus*.

There are few characteristics worthy of special notice with regard to the anatomical features of *C. vesiculosus* and *C. bullatus*; they are fundamentally similar to those of ordinary species of *Chlamys* and *Aequipecten*. Tentacles, though not numerous, are considerably long. Eyes arranged at irregular intervals along the edge of the mantle are proportionally large, presumably reflecting deep (dark) habitat.

Of the above-mentioned living materials of *C. vesiculosus*, all the relatively large-sized individuals (over 13 mm in length or height) in the July and August subsamples, irrespective of phenotype and the presence or not of growth ring(s) on the surface, have a large swollen crescentic gonad attached to the adductor muscle. Its proximal part is creamy and certainly represents the male reproductive gland (testis), whereas its distal part is

24

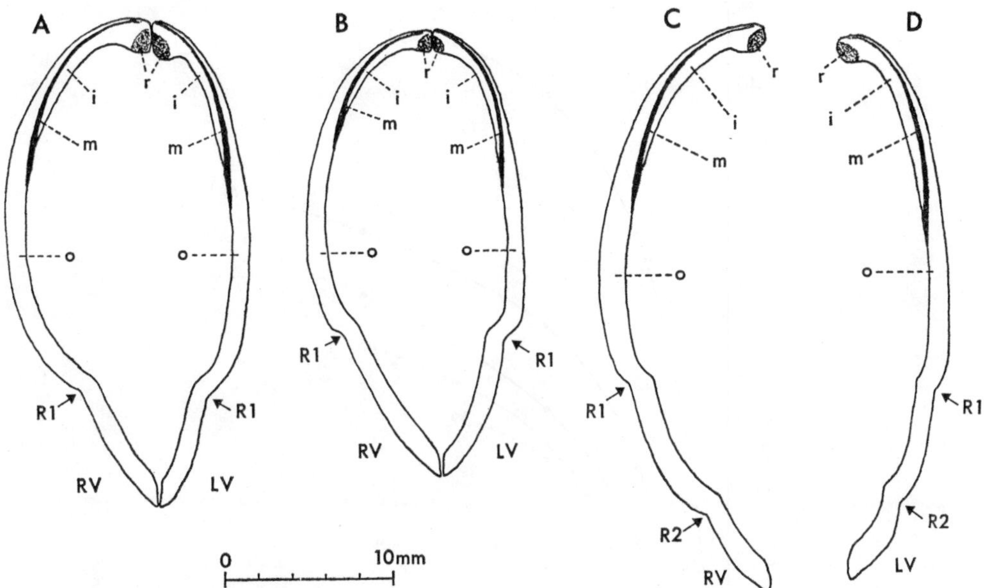

Fig. 6. Subvertical sections of some adult specimens of *Cryptopecten vesiculosus*, showing significant stepwise growth rings. *A*: conjoined specimen belonging to sample
Is (UMUT RM16138e), phenotype *R*; B: conjoined specimen belonging to sample
Is (UMUT RM16138f), Phenotype *Q*; C: Right valve of sample *Ab 1* (UMUT
CM16051h), Phenotype *R*; D: Left valve belonging to sample *Ab 1* (UMUT CM
16051i), Phenotype *R*. RV: right valve, LV: left valve, r: resilifer, i: inner shell layer
(foliated calcite), m: myostracum (aragonite), o: outer shell layer (foliated calcite), R1:
first growth ring, R2: second growth ring.

orange-vermilion and contains the female reproductive gland (ovary) (see Frontispiece).
The boundary between the two glands is irregular but generally sharp. Both reproductive
glands are also developed in some large individuals of the May subsamples, but not in
any of those of the March, October and November ones. The spawning season must
therefore be mainly in summer (or early autumn). It is likely that, as commonly seen in
other hermaphroditic bivalves, the male gland becomes mature prior to the female gland
within the same individual, so that self-fertilization is avoided.

Most Pleistocene, Holocene and Recent large individuals of *C. vesiculosus* from Japan
(over 20 mm in length and height) have one or more strong growth rings on the surface.
The rings are often so conspicuous that the shells appear to be constricted, as seen in the
profiles (Plates 5 and 6) and cross sections (Fig. 6). The curvature of shell surface, both
external and internal, commonly changes drastically at the ring; it is strongest just before
the formation of each ring and becomes suddenly much weaker after the formation.
Judging from the periodicity of growth increments in the outer shell layer, it is inferred
that rings are laid at times of unfavorable growth conditions. In many fossil and Recent
large individuals several (five at the maximum) rings are clearly observed, and the intervals become regularly narrower with growth. The distance from the origin of growth to
each ring seems to agree well with a logistic curve (several examples are shown in Fig. 7),
which may be a general pattern of absolute growth (Simpson, Roe and Lewontin, 1960,

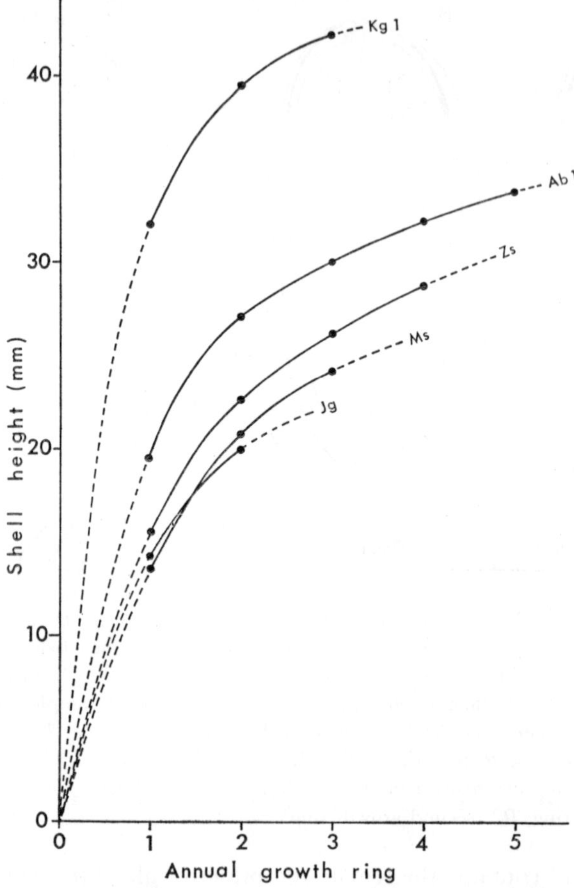

Fig. 7. Logistic curves of absolute growth inferred from the growth rings in some
selected individuals belonging to *Cryptopecten vesiculosus*. Kg 1: left valve (UMUT
CM16033a) of sample *Kg 1*, Ab 1: left valve (UMUT CM16051g) of sample *Ab 1*,
Zs: conjoined valves (UMUT RM16083a) of subsample *Zs (4)*, Ms: left valve (UMUT
CM16057g) of sample *Ms*, Jg: conjoined valves (UMUT RM16061b) of subsample
Jg (2).

fig. 49). Though the growth pattern before the formation of the first ring (or sigmoidal
part of a logistic curve) is actually unknown, periodical formation of growth rings is also
strongly suggested from the biometrical ground.

At first I presumed that these rings were formed in winter, simply because *C. vesiculosus* is a warm-current species. Contrary to my expectation however, as a result of my
examination on living populations in Sagami Bay, it was found that a growth ring is
formed between early August and early October. More precisely, summer specimens
with fully developed reproductive grands always show strongly inflated ventral periphery. It is therefore reasonable to consider that the increased internal space produced
by the unusually strong curvature corresponds to the full development of the gonad. I
presume further that in this season shell growth ceases or is much retarded in relation
to minimized metabolism. As seen in most large individuals of the October subsamples,

the quantity of soft part greatly decreases in autumn. In October very small individuals, which probably represent a new generation, are found together with large individuals exhibiting slight shell growth begun after the completion of the last ring.

The present observation about the formation of growth rings seems to be consistent with the morphological features in several other dated living samples [*Kz, Ma, Aj (1, 2), Os, Zs (1, 3, 4, 8), Hs (2), Ts (2, 3), Hk, Mb*], which were collected at several stations along the Pacific coast of central Honshu in various seasons. Among these are three small samples were obtained in summer from Zenisu, a deep-water bank (ca. 100 m) southwest of Kôzu Island. Their soft parts being completely dried, I was unable to observe the development of reproductive glands, but it is noticed that the individuals of *Zs (3, 4)* collected on July 10, 1969 show strong curvature of shell near the ventral margin indicating the stage during or just before the formation of a growth ring. On the other hand, the individuals of *Zs (8)* collected on August 3, 1971 have a bit of added shell after the last ring. In an individual of *Hs (2)*, which was collected on July 12, 1967 on another bank near Zenisu, however, slight shell growth already began after the formation of the last ring. Judging from the relation between the collecting dates and the growth amount after the last ring, spawning is generally considered to occur mainly in summer in sea areas other than Sagami Bay as well, but the season may vary slightly with the area, year and individual. It is quite likely that the reproductive season is somewhat earlier in southern populations, provided that spawning occurs under impetus from an increase in water temperature. Incidentally, the breeding seasons of such commercially important pectinids as *Pecten (Notovola) albicans, Patinopecten (Mizuhopecten) yessoensis* and *Chlamys farreri nipponensis* are considerably different in the various sea areas around Japan (Yoshida, 1964; etc.).

The present data are still insufficient to permit reconstruction of life cycle and population structure. Since 10 mm mesh nets were used for dredging, size distribution of smaller individuals is difficult to determine. In July and August subsamples well-developed gonads are observed not only in the individuals with growth ring(s) but also in a considerable number of medium-sized individuals without a ring. These individuals are, however, accompanied by still smaller ones without any developed gonad or any inflation of ventral periphery. Therefore it can be most likely, even though not conclusive, that most individuals reach sexual maturity in the second summer. The distance from the hinge-line to the first growth ring (H_1) varies greatly within each sample. This may be due to a long-term breeding season, but further investigation by regular sampling may be needed, since, as discussed later (p. 36), the frequency distribution of H_1 is sometimes clearly bimodal.

The seasonality of spawning as well as the formation of growth rings has been intensively studied in *Pecten (Pecten) maximus* in the Irish Sea. According to Mason (1957), who examined regular samples collected at weekly or fortnightly intervals over two years, the scallop grows from spring to December and ceases growing in winter. Growth rings are laid down annually in early spring when water temperature is lowest. The spawning season of this species in this sea area is quite long, being mostly in late August or September, but there is also less common spawning in the period between April and early August. Mason's observation seems, therefore, to indicate that the formation of growth rings is unrelated to the breeding season, an indication, however, which is somewhat in

28

contradiction to Ree's (1957) interpretation. Mason (1957) cited his 1953 thesis: "Spawning cannot possibly be responsible for the cessations of growth in the animal's first two winters, since scallops do not spawn for the first time until the summer after the deposition of the second growth-rings." Because *C. vesiculosus* has much smaller shell size and much fewer growth rings than *P.* (*P.*) *maximus*, it is only natural to assume that sexual maturity is attained somewhat earlier.

Growth rings which I have observed in several specimens of *P.* (*P.*) *maximus* preserved in our museum and several other institutions, are not accompanied by periodical change of surface convexity. This is also the case with many other pectinids. Such rings, I presume, are not related to breeding. However, remarkably stepwise growth rings, strikingly similar to those of *C. vesiculosus*, are frequently met with in *Chlamys* (*Swiftopecten*) *swiftii* (Bernardi, 1858) and *Nodipecten nodosus* (Linnaeus, 1758). Among Japanese fossil species such remarkable growth rings appear in *Chlamys cosibensis* (Yokoyama, 1911) and its relatives from the Miocene—Pleistocene (see Masuda, 1959). The origin of stepwise growth rings should be more intensively studied in each species, but analogy with *C. vesiculosus* leads me to suspect that they are primarily related to the development of gonads and temporal cessation of growth after annual spawning.

In bivalves growth rings undoubtedly arise from various causes. In *C. vesiculosus*, however, it is concluded that in adult individuals they are laid down annually, mainly in summer or early autumn, only during the season of reproductive activity and also that their number indicates the lowest possible age of each individual. This conclusion may be applicable to fossil samples of this species, e. g. in comparison of age composition between samples and the evaluation of seasonal heterogeneity.

The relative abundance of small individuals without any growth ring in the sample *Jg* (*1–26*) seems to indicate a high mortality rate in the immature stage. Since the sample was obtained using nets of 10 mm mesh, there is doubtless a greater abundance of immature individuals than the number collected would indicate. Of 265 living individuals in this sample only 22 have only one growth ring. Also noteworthy is the fact that many dead shells of medium size, between 12 mm and 20 mm, show strongly convex ventral periphery, suggesting that many mature individuals passed away just after the first spawning, only a fraction surviving to join in reproductive activity in the following summer.

This may not necessarily be true however in other living and fossil populations. Large specimens with more than two rings are quite common in many Pleistocene fossil samples from Boso Peninsula (e. g. samples *Sn 1–3*, *Ny 1–4*, *Jz*, *Ic* and *Ab 1–2*) as well as in some Recent ones (e. g. *Zs*, *Ts* and *Az*).

For example, as shown in Text-fig. 9, of 912 collected left valves of fossil sample *Jz*, 740, 487, 231, 34 and 2 individuals possess one, two, three, four and five growth rings, respectively. No significant difference exists between the growth patterns of the two phenotypes. Because of the sampling bias and post-mortem sorting, these figures do not seem to be directly indicative of mortality rate, but it is evident that the average number of growth rings is generally much larger in comparison with Recent population samples. The mortality rate and average span of one generation may vary considerably among local populations in time and space.

In most Pliocene samples (e. g., *Sh 1–2*, *Nj*, *Ik* and *Kg 1–2*) growth rings, if present, are comparatively weak (Plate 7, Figs. 9–12). It is, however, doubtful whether the difference is entirely attributable to phyletic change, because weak rings are also common in some Recent samples from the southern East China Sea (e. g., *Ec*, *My*)

Also noteworthy is the fact that several specimens (about 13 mm in length and height) of *C. bullatus*, which were collected on June 15, 1983 off the coast of Shirahama of Kii Peninsula (sample *Sr* (*2*, *3*) (*B*)), possess a considerably developed gonad (see Table 2 and Frontispiece). An associated specimen of *C. vesiculosus* (sample *Sr*) however, which was obtained by the same dredge as *Sr* (*2*) (*B*), does not show any developed gonad, though its shell size is considerably large. The reproductive seasons of the two species are probably different, at least in this sea area.

Large specimens of *C. bullatus*, *C. yanagawaensis* and *C. spinosus* often have one or two stepwise growth rings, which are, however, less conspicuous than those of *C. vesiculosus*. In *C. nux* growth ring(s) are rarely encountered even in such large population samples as *Kk* (*N*) and *Bh* (*N*). Judging from the small ultimate size and occasionally strongly inflated ventral part, I presume that almost all the mature individuals of this species pass away immediately after the first spawning.

CHAPTER 7

Intrapopulational Variation and Relative Growth

As was fully discussed and enumerated by Mayr (1963, pp. 138–163; 1969, pp. 144–163), individual variation within a single interbreeding population may be caused by various genetic and nongenetic factors. Because environmentally induced morphological modification has almost nothing to do with evolutionary change, causal evaluation of variation, though not necessarily easy in fossil and marine molluscs, is of primary importance for evolutionary studies of this kind.

Individual variation of morphological characters are classifiable, though only for convenience, into two types: continuous and discontinuous. Various morphometric (including meristic) characters such as body size at a definite growth stage, form ratios between two linear measurements, angles, and number of serially arranged organs commonly vary continuously, and usually show histograms resembling normal frequency distribution. This continuously varied unit character is generally regarded as being controlled by a large number of genetic and environmental factors. Discontinuous variation, on the other hand, is thought to be controlled by a single or a few genetic or nongenetic factor(s). Both types of individual variation are seen in each large population of *Cryptopecten*.

Since *Cryptopecten bullatus*, *Cryptopecten nux* and extinct species of this genus are represented by relatively small samples in the present collection, the detailed examination of intrapopulational variation is for the most part restricted to several large samples of *C. vesiculosus*. The size of some of the samples precluded measurement of all the specimens, so 50 right and 50 left (or 50 conjoined) valves were randomly selected. In Middle Pleistocene and later samples two discrete phenotypes are distinguishable by the sculpture of the disk, and they were treated separately. Computations for statistics were made on these reduced samples excluding some broken valves which were inadequate for measurement. Meristic variation was examined on these and several more samples. Phenotypic frequency of dimorphism however, was examined on the basis of all individuals in every fossil and Recent sample.

A. Continuous Variation

In the quantification of continuous variation, a basic and serious problem is that of how to eliminate various influences caused by the age heterogeneity. Needless to say, all the samples representing local populations consist of individuals of different average size, and none of the samples show a definite pattern of size distribution. Furthermore, in dead shells and fossils the size distribution may be greatly influenced by differential sorting during post-mortem transportation and deposition. Various biases in shell size are also inevitable during sampling in the field and preparation in the laboratory. For example, the representation of small-sized individuals is inevitably influenced by the

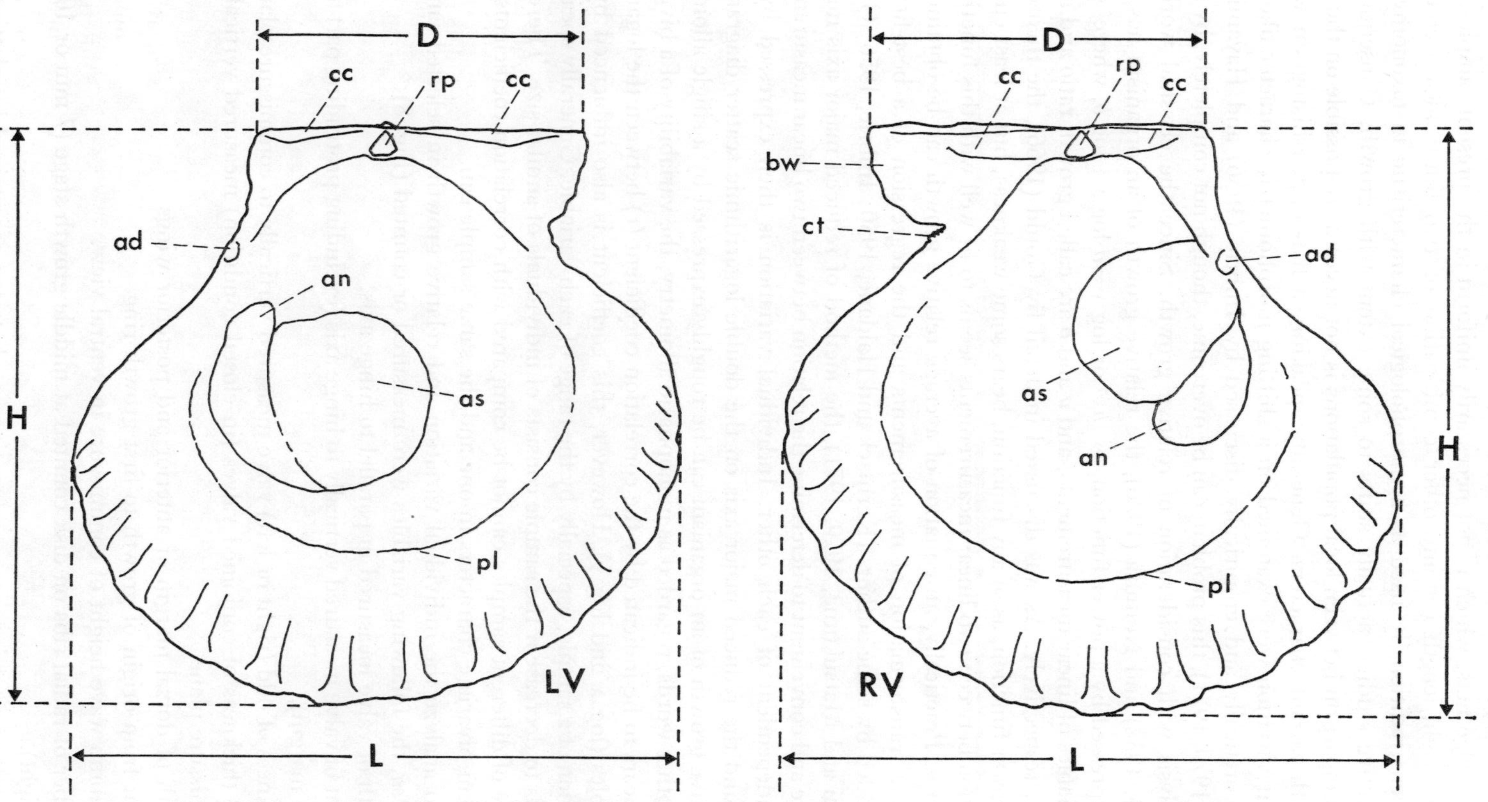

Fig. 8. Internal morphology of *Cryptopecten vesiculosus*, showing linear measurements. L: length, H: height, D: length of dorsal margin of two wings, ad: auricular denticle, an: adductor muscle scar (non-striate portion), as: adductor muscle scar (striate portion), bw: byssal wing, cc: cardinal crura, ct: denticles of ctenolium, pl: pallial line, rp: resilial pit.

32

mesh size of dredge nets, which is not necessarily uniform in the present samples. The form ratio of height/length (or any other combination of two linear measurements), which has been traditionally used as a morphological characteristic in taxonomy and other comparative studies, actually shifts to some extent with growth. Consequently, meaningful comparison between local populations is not necessarily possible on the basis of such growth-variant characters. Generally speaking, a flat-topped histogram would be obtained, if variation were examined on a shifting morphometric character like this.

As was theoretically and empirically discussed by Imbrie (1956) and Hayami and Matsukuma (1970, 1971), this problem can be overcome, though not completely, by some bivariate analysis with consideration of relative growth. Since the classical works of Huxley (1924, 1932) and Nomura (1926), the relative growth of an organism has been commonly expressed by a power function, $y=bx^a$ or $\log y=a \log x+\log b$, where x and y are two variables of linear measurements, and a and b are called growth ratio and initial growth index, respectively. As was discussed in detail by Gould (1966), the theoretical basis of the power function, as of any function, bears some weakness, but at least empirically the relation between two linear measurements seems to fit well with this function in many organisms. Practically, an equation of average relative growth can be obtained by the logarithmic transformation of measurements and the regression of a best-fit line. As recommended by some authors (Kermack and Haldane, 1950; Imbrie, 1956; Gould, 1966; Hayami and Matsukuma, 1970, 1971), the method of reduced major axis may be most adequate and convenient to determine the relation between two linear measurements which are independent of each other. Individual variation is then expressed by the dispersal around the reduced major axis on the double logarithmic scatter diagram.

If the relative growth of an organism can be roughly expressed by a single allometric equation, in other words, regarded as monophasic allometry, the variability of a bivariate character appears to be indicated by the correlation coefficient (r) between the logarithm of two variables ($\log x$ and $\log y$). However, this coefficient is also influenced by the size composition of a sample, especially by the range of each variable. Generally speaking, its value tends to decrease if the sample consists of individuals of similar size. Therefore, the variability of different samples cannot be compared with correlation coefficients, but that of different bivariate characters in one and the same sample can.

In order to analyze the individual variation and relative growth in selected samples of *Cryptopecten*, the following variables were measured or counted (Fig. 8):

L: Length of valve measured in parallel to hinge axis.

H: Height of valve measured vertically to hinge axis (excluding protruding part above hinge margin).

T: Thickness of odd (right or left) valve measured vertically to commissure plane.

C: Total thickness of conjoined valves (in closed condition) measured vertically to commissure plane.

D: Length of dorsal margin of anterior and posterior wings.

H_1: Height from origin of growth to first growth ring.

W: Maximum wave height of commissure in ventral view.

X: Number of radial ribs on disk counted at middle growth stage (7 mm or 10 mm in height).

The adequacy of applying a power function, $y=bx^a$, was ascertained by plotting the

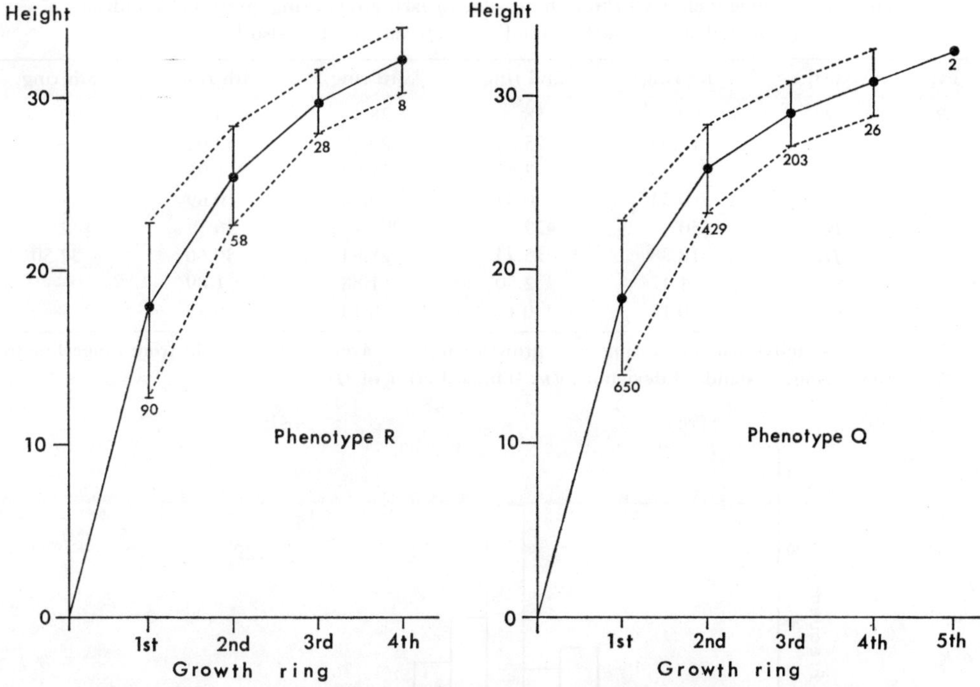

Fig. 9. Patterns of absolute growth in two phenotypes of *Cryptopecten vesiculosus*, inferred from the shell height from the origin of growth to each stepwise growth ring. Sample *Jz* (only left valves). Vertical bars and numerals below indicate standard deviations and sample sizes, respectively.

bivariate data on double logarithmic papers (some examples shown in Figs. 9, 10, 13).

The result of bivariate analysis is shown in Tables 4–10, where the following abbreviations are used:

N: Number of examined individuals

r: Correlation coefficient between two variables

a: Growth ratio (slope of reduced major axis in double logarithmic scatter diagram), which is obtained by $a = s_{\log y}/s_{\log x}$ [s: standard deviation]

σ_a: Standard error of growth ratio, which is obtained by $\sigma_a = a\sqrt{(1-r^2)/N}$

b: Initial growth index (anti-logarithm of y-intercept of reduced major axis in double logarithmic scatter diagram), which is obtained by $\log b = \log y - a \log x$

RG: Mode of relative growth

I: "Isometry" (null hypothesis of $a=1$ not rejected with 95 percent confidence)

NA: Negative allometry (a significantly smaller than 1 with 95 percent confidence)

PA: Positive allometry (a significantly larger than 1 with 95 percent confidence)

In the same tables three values of form ratios are also indicated. They were theoretically obtained from each allometric equation at three fixed sizes ($L=10$ mm, 20 mm and 30 mm). The values in parentheses indicate that the observed range of L does not cover the fixed size.

Generally speaking, comparison of the growth ratios (a) of allometric equations seems

Table 3. Average shell height from hinge-line to each growth ring, presumably indicating annual shell growth. Sample Jz (left valve). See also Fig. 9.

Type	Statistics	1st ring	2nd ring	3rd ring	4th ring	5th ring
R	N	90	58	28	8	0
	$\bar{H}i$	17.63	25.39	29.62	32.04	—
	s	5.03	2.85	1.92	1.95	—
	σ_H	0.53	0.37	0.36	0.69	—
Q	N	650	429	203	26	2
	$\bar{H}i$	18.38	25.74	28.84	30.66	32.50
	s	4.47	2.50	1.88	1.89	—
	σ_H	0.18	0.12	0.13	0.37	—

N: number of individuals possessing each growth ring; $\bar{H}i$: average shell height from hinge-line to each growth ring; s: standard deviation; σ_H: standard error of $\bar{H}i$.

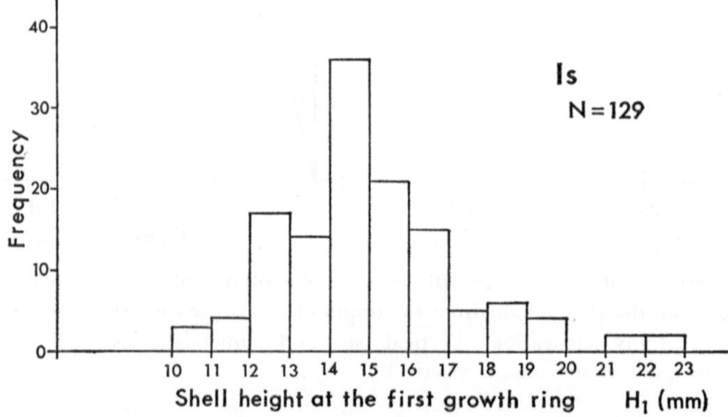

Fig. 10. Example of the frequency distribution of H_1 (shell height at the first growth ring). Sample Is of *Cryptopecten vesiculosus*.

to be meaningful, but that of the initial growth indices (b) does not because the latter means the value of y at $x=1$ and is clearly dependent on a (see Imbrie, 1956, for the statistical comparison of a bivariate character between two samples). The form ratios at fixed sizes are probably useful for comparative studies, because the influence of size difference between samples is largely eliminated.

1. Shell Size and Growth Rate [Table 3]

Shell size is undoubtedly one of the most easily measurable univariate characters, variation being controlled primarily by polygenic factors. The simple overall size of shell is, however, not very useful, unless the influence of age heterogeneity is eliminated by some adequate method. In this respect, the partial height of shell from the origin of growth (or hinge-axis) to the first stepwise growth ring, here defined as H_1, may be more significant, because it seems to indicate the growth amount from fertilization to the first unfavorable season for growth. Judging from the data on living populations in the eastern part of Sagami Bay, this value directly indicates size at attainment of sexual maturity. Annual growth amount of an individual seems to be shown by the interval of successive

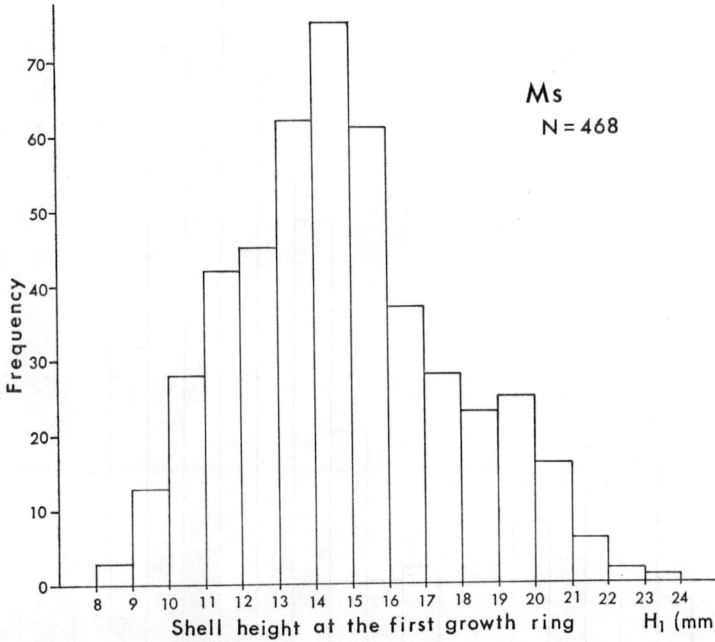

Fig. 11.   Example of the frequency distribution of H_1 (shell height at the first growth ring). Sample *Ms* of *Cryptopecten vesiculosus*.

growth rings, each of which is laid in relation to the annual reproductive season (see p. 26).

The intrapopulational variation of H_1 is quite wide in every sample, the coefficient of variation (V) commonly exceeding 15 (Table 13). As exemplified by sample *Jz* (Table 3, Fig. 9), however, the variation generally becomes much narrower at the subsequent ring(s). The reason why the size variation decreases with growth must be further investigated, but it is, I think, mainly due to the long duration of breeding season. If fertilization occurred for several successive months every year, the reduction of variation would be logically explicable, because the influence of age heterogeneity and the absolute growth rate are generally much reduced with growth.

In most Recent samples the frequency distribution of H_1 is unimodal, the mode lying at about 15 mm (Figs. 10 and 11). In some Late Pleistocene samples, however, the distribution is clearly bimodal. In the fossil sample *Jz*, for example, there are two clusters in the histogram of H_1; the mode of the smaller cluster lies at about 10 mm and that of the larger one is near 21 mm (Fig. 12).

Bimodal distribution of H_1 may in fact be due to the occurrence of two breeding seasons a year, as is the case with *Pecten (Pecten) maximus* in the Irish Sea (Mason, 1957). In the present case, however, I presume for the following reasons that the first ring in the smaller cluster is largely unrelated to reproductive activity.

1)   The first ring in the smaller cluster is generally, if not discriminately, weaker and less stepwise in comparison with subsequent rings.

2)   The growth rate and ultimate shell size in these Late Pleistocene samples are

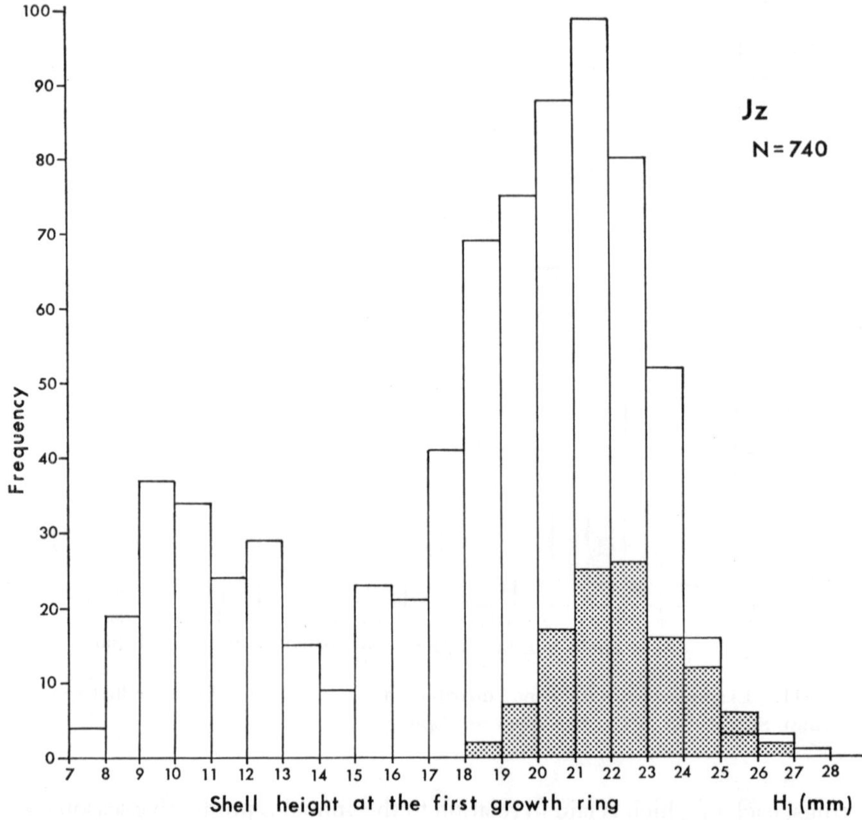

Fig. 12. Example of the frequency distribution of H_1 (shell height at the first growth ring). Sample Jz (left valves only) of *Cryptopecten vesiculosus*. Shadowed histogram indicates the distribution of H_2 (shell height at the second ring) in the individuals belonging to the smaller cluster ($H_1 < 14$ mm).

much larger than those of Recent ones. Therefore it is rather unlikely that individuals reach sexual maturity at such a small size.

3) The average shell size of the smaller cluster at the next (second) ring is not much different from that of the larger cluster at the first ring (Fig. 12).

If this interpretation is correct some amendments may be necessary as to the numbering of growth rings and statistical treatment of shell size. The average of H_1 would then becomes considerably larger, and its standard deviation somewhat smaller. Furthermore, not only in the bimodal samples but also in apparently unimodal samples, it is by no means proven that the first ring at very small size (less than 10 mm) indicates sexual maturity. It can be said, however, that this ambiguous ring occurs only in a small fraction of each population sample. In other words, even if the numbering of rings were partly wrong, this would not seriously influence the discussion and conclusion at this level.

2. Length [*L*] versus Height [*H*]   [Tables 4 and 5]

The simple ratio of H/L appears to indicate the degree of vertical elongation of disk.

Table 4. Allometric relation between L and H, $H=bL^a$ (Right valves).

Sample	Pheno-type	N	r	$a\pm\sigma_a$	b	RG	Form ratio H/L		
							$L=10$ mm	$L=20$ mm	$L=30$ mm
Sh 1	Q	7	0.9985	0.9072 ± 0.0188	1.3021	NA	1.052	0.986	(0.950)
Ik	Q	50	0.9963	0.9261 ± 0.0113	1.2088	NA	(1.020)	0.969	0.940
Ob	Q	26	0.9962	0.8990 ± 0.0154	1.3920	NA	(1.103)	1.029	0.987
Tm 1	Q	9	0.9823	0.8468 ± 0.0529	1.6712	NA	(1.174)	1.056	(0.993)
Ab 1	Q	50	0.9954	0.9298 ± 0.0126	1.2278	NA	(1.045)	0.995	0.967
	R	42	0.9959	0.9351 ± 0.0131	1.2399	NA	(1.068)	1.021	0.994
Ms	Q	50	0.9949	0.9102 ± 0.0130	1.3085	NA	(1.064)	1.000	0.964
	R	50	0.9960	0.9427 ± 0.0119	1.2020	NA	(1.053)	1.012	0.989
Hy	Q	18	0.9992	0.9475 ± 0.0089	1.1671	NA	1.034	0.997	(0.976)
	R	20	0.9990	0.9723 ± 0.0097	1.1234	NA	1.053	1.033	(1.022)
Jg (8)	Q	50	0.9968	0.9234 ± 0.0104	1.2641	NA	(1.060)	1.005	(0.974)
	R	50	0.9960	0.9318 ± 0.0118	1.2517	NA	(1.070)	1.020	(0.993)
Su (29)	Q	44	0.9957	0.9163 ± 0.0123	1.2788	NA	(1.055)	0.995	(0.962)
	R	32	0.9967	0.9349 ± 0.0134	1.2273	NA	1.056	1.010	(0.984)
Is	Q	50	0.9945	0.9470 ± 0.0140	1.1687	NA	(1.034)	0.997	(0.976)
	R	50	0.9950	0.9633 ± 0.0143	1.1290	NA	(1.038)	1.011	(0.997)
Ec (2)	Q	40	0.9982	0.9132 ± 0.0087	1.3002	NA	1.065	1.002	(0.968)
	R	17	0.9971	0.9018 ± 0.0166	1.3379	NA	1.067	0.997	(0.958)
Ta (4)	Q	50	0.9987	0.9443 ± 0.0068	1.1860	NA	1.043	1.004	(0.981)
	R	28	0.9975	0.9478 ± 0.0127	1.1777	NA	1.044	1.007	(0.986)
Hm	Q	21	0.9977	0.8945 ± 0.0132	1.3781	NA	(1.081)	1.005	0.963
	R	8	0.9987	0.9130 ± 0.0165	1.3134	NA	(1.074)	1.012	0.977
Kk (S)	—	39	0.9979	0.9397 ± 0.0097	1.1953	NA	1.040	0.998	(0.974)
Kk (B)	—	19	0.9955	0.9480 ± 0.0206	1.1207	NA	0.994	(0.959)	(0.939)
Kk (N)	—	50	0.9943	0.9347 ± 0.0141	1.1535	NA	0.992	(0.949)	(0.924)
Sm (N)	—	6	0.9975	1.0129 ± 0.0295	1.0032	I	1.034	(1.043)	(1.048)
Bh (N)	—	19	0.9966	0.9182 ± 0.0174	1.2011	NA	0.995	(0.940)	(0.909)

In almost all the samples of *C. vesiculosus* this ratio may or may not exceed 1, and varies considerably among individuals. The frequency distribution of H/L, as well as other form ratios, is commonly more or less flat-topped (platykurtic).

For instance, various basic statistics of this character calculated on 100 individuals randomly selected from the sample *Is* are as follows:

Sample size: $N=100$
Arithmetic mean of H/L: $\bar{x}=1.003\pm0.002$
Standard deviation of H/L: $s=0.024$
Coefficient of variation: $V=2.350$
Observed range: $O.\,R.=0.949-1.056$

$$\text{Kurtosis:}\ Ks=\frac{\sum\limits_{i=1}^{100}(x_i-\bar{x})^4}{Ns^4}-3=-0.536$$

The kurtosis is an interesting index. As was clearly explained by Simpson, Roe and Lewontin (1960), a negative (or positive) value of Ks indicates that the frequency distribution is platykurtic (or leptokurtic) in comparison with ideal normal distribution. The

Table 5. Allometric relation between L and H, $H=bL^a$ (Left valves).

Sample	Pheno-type	N	r	$a\pm\sigma_a$	b	RG	Form ratio H/L		
							$L=10$mm	$L=20$mm	$L=30$mm
Sh 1	Q	9	0.9949	0.9480±0.0319	1.1703	I	(1.038)	1.001	(0.981)
Ik	Q	50	0.9961	0.9002±0.0112	1.3106	NA	(1.042)	0.972	0.933
Ob	Q	21	0.9975	0.9155±0.0141	1.3380	NA	1.101	1.039	(1.004)
Tm 1	Q	8	0.9966	0.9288±0.0271	1.2925	NA	1.097	1.044	1.015
Ab 1	Q	50	0.9980	0.9138±0.0082	1.2991	NA	1.065	1.003	0.969
	R	24	0.9957	0.9074±0.0172	1.3548	NA	(1.095)	1.027	0.989
Ms	Q	50	0.9972	0.9334±0.0099	1.2289	NA	1.054	1.007	0.980
	R	50	0.9959	0.9152±0.0117	1.2909	NA	(1.061)	1.001	(0.967)
Hy	Q	18	0.9992	0.9475±0.0089	1.1671	NA	1.034	0.997	(0.976)
	R	20	0.9990	0.9723±0.0097	1.1234	NA	1.054	1.034	(1.022)
Jg (8)	Q	50	0.9961	0.9347±0.0120	1.2315	NA	(1.060)	1.013	(0.986)
	R	50	0.9963	0.9353±0.0114	1.2426	NA	(1.071)	1.024	(0.997)
Su (29)	Q	43	0.9967	0.9194±0.0114	1.2780	NA	1.062	1.004	(0.972)
	R	26	0.9982	0.9294±0.0109	1.2489	NA	1.062	1.011	(0.982)
Is	Q	50	0.9945	0.9470±0.0140	1.1687	NA	(1.034)	0.997	(0.976)
	R	50	0.9950	0.9633±0.0143	1.1290	NA	(1.037)	1.011	(0.997)
Ec (2)	Q	24	0.9986	0.8725±0.0094	1.4567	NA	1.086	0.994	(0.944)
	R	16	0.9986	0.9137±0.0121	1.3033	NA	1.068	1.006	(0.972)
Ta (4)	Q	50	0.9987	0.9212±0.0066	1.2580	NA	1.049	0.993	(0.962)
	R	30	0.9987	0.9473±0.0088	1.1827	NA	1.048	1.010	(0.989)
Hm	Q	19	0.9966	0.9094±0.0172	1.3298	NA	(1.079)	1.014	(0.977)
	R	8	0.9976	0.9025±0.0221	1.3559	NA	(1.083)	1.012	(0.973)
Kk (S)	—	25	0.9986	0.9439±0.0100	1.1795	NA	1.036	0.997	(0.975)
Kk (B)	—	19	0.9956	0.9250±0.0199	1.1781	NA	0.991	(0.941)	(0.913)
Kk (N)	—	50	0.9951	0.9346±0.0131	1.1561	NA	0.994	(0.950)	(0.926)
Sm (N)	—	6	0.9964	0.9224±0.0317	1.2441	NA	1.041	(0.986)	(0.956)
Bh (N)	—	19	0.9966	0.9182±0.0174	1.2011	NA	0.995	(0.940)	(0.909)

negative value of *Ks* in this case is rather remarkable, indicating a considerably flat-topped distribution. This pattern is perhaps largely due to the age heterogeneity, because the average value of H/L actually becomes much smaller with increase of shell size, as shown below:

Size range (in mm)	Number of individuals	Mean of H/L (with standard error)
$12.5 \geqq L \geqq 14.9$	7	1.021±0.007
$15.0 \geqq L \geqq 17.4$	10	1.018±0.006
$17.5 \geqq L \geqq 19.9$	23	1.007±0.004
$20.0 \geqq L \geqq 22.4$	22	1.004±0.005
$22.5 \geqq L \geqq 24.9$	13	0.989±0.006
$25.0 \geqq L \geqq 27.4$	22	0.996±0.005
$27.5 \geqq L \geqq 29.9$	3	0.971±0.014
Total	100	1.003±0.002

The average relative growth between length (L) and height (H) was examined on 44 lots of 13 selected samples of *C. vesiculosus* by the method of reduced major axis. In all

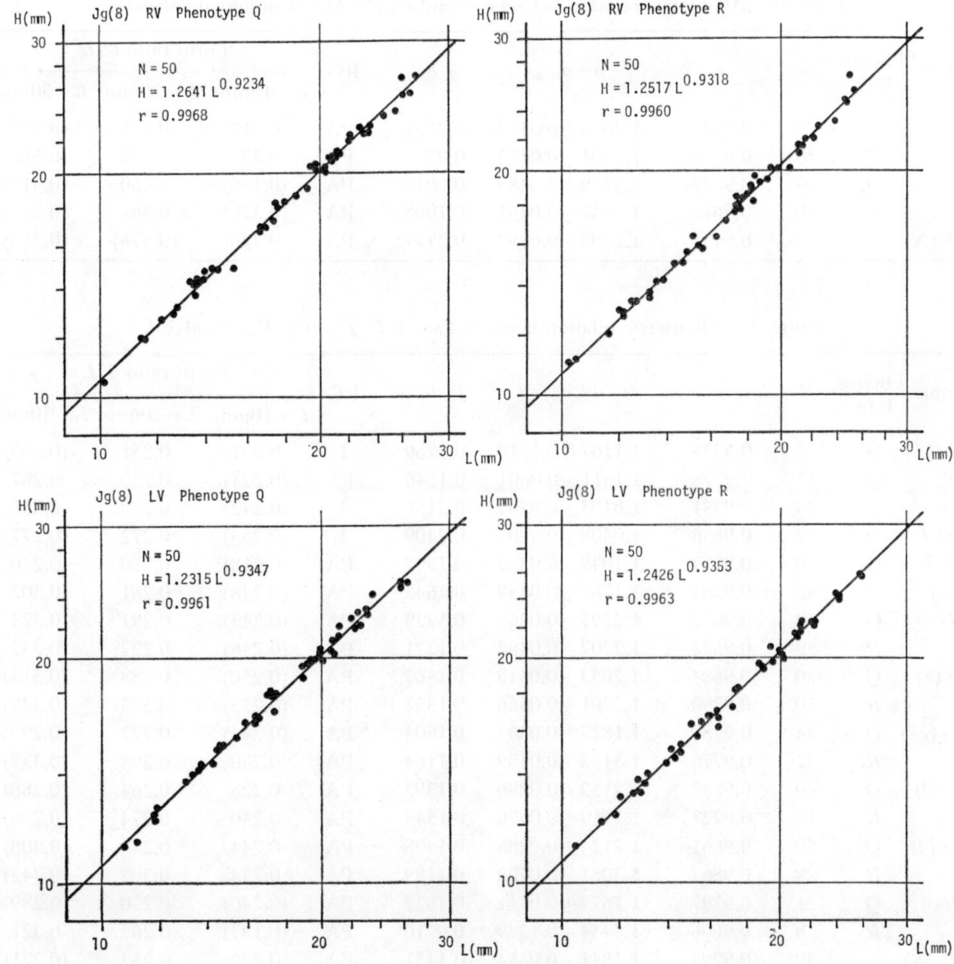

Fig. 13.   Example of the allometric relation between L (length) and H (height). Sample *Jg (8)* of *Cryptopecten vesiculosus*. RV: right valves, LV: left valves. The high correlation coefficients and the reduced major axes on double logarithmic scatter diagrams indicate that H is significantly negatively allometric to L.

of these samples, both in right and left valves, log L shows extremely high correlation with log H; the correlation coefficient (r) exceeds 0.99 except for one small sample. As an example, the L-H relation in subsample *Jg (8)* is shown in Fig. 13. The growth ratio (a) varies considerably with the sample, but in almost all lots the values are significantly smaller than 1, and the relative growth is regarded as negatively allometric with 95 percent confidence (see Hayami and Matsukuma, 1970, 1971, for the statistical test for discrimination between "isometry" and allometry). The obtained values of a are seldom significantly different between two valves and between two phenotypes within one and the same sample, though they sometimes differ between different samples.

C. spinosus sp. nov., which is proposed in this article on fossil sample *Kk (S)*, seems to exhibit the same tendency with respect to this bivariate character as *C. vesiculosus*.

Table 6. Allometric relation between L and C, $C=bL^a$ (Conjoined valves).

Sample	Pheno-type	N	r	$a \pm \sigma_a$	b	RG	Form ratio C/L		
							$L=10$mm	$L=20$mm	$L=30$mm
Hy	Q	18	0.9836	1.2075 ± 0.0513	0.2551	PA	0.411	0.475	(0.517)
	R	20	0.9863	1.2801 ± 0.0472	0.2285	PA	0.435	0.529	(0.592)
Is	Q	50	0.9627	1.2759 ± 0.0488	0.2014	PA	(0.380)	0.460	(0.515)
	R	50	0.9678	1.2945 ± 0.0461	0.1908	PA	(0.376)	0.461	(0.520)
Bh (*N*)	—	19	0.9442	1.1805 ± 0.0892	0.2774	PA	0.420	(0.476)	(0.513)

Table 7. Allometric relation between L and T, $T=bL^a$ (Right valves).

Sample	Pheno-type	N	r	$a \pm \sigma_a$	b	RG	Form ratio T/L		
							$L=10$mm	$L=20$mm	$L=30$mm
Sh 1	Q	7	0.9338	1.1167 ± 0.1510	0.1769	I	0.231	0.251	(0.263)
Ik	Q	45	0.9378	1.1611 ± 0.0601	0.1546	PA	(0.224)	0.250	0.267
Ob	Q	26	0.9581	1.0493 ± 0.0589	0.2160	I	(0.242)	0.250	0.255
Tm 1	Q	9	0.9656	1.0408 ± 0.0902	0.2409	I	(0.265)	0.272	(0.277)
Ab 1	Q	50	0.9779	1.1942 ± 0.0353	0.1397	PA	(0.218)	0.250	0.270
	R	42	0.9704	1.1787 ± 0.0439	0.1643	PA	(0.248)	0.281	0.302
Ms	Q	50	0.9652	1.2595 ± 0.0466	0.1339	PA	(0.243)	0.291	0.324
	R	50	0.9633	1.2707 ± 0.0482	0.1321	PA	(0.246)	0.297	0.332
Jg (*8*)	Q	50	0.9694	1.2033 ± 0.0418	0.1567	PA	(0.250)	0.288	(0.313)
	R	50	0.9790	1.2701 ± 0.0366	0.1351	PA	(0.252)	0.303	(0.339)
Su (*29*)	Q	44	0.9789	1.1827 ± 0.0364	0.1604	PA	(0.244)	0.277	(0.299)
	R	32	0.9776	1.3143 ± 0.0489	0.1164	PA	0.240	0.298	(0.339)
Ec (*2*)	Q	40	0.9888	1.2132 ± 0.0286	0.1393	PA	0.228	0.264	(0.288)
	R	17	0.9723	1.1919 ± 0.0676	0.1543	PA	0.240	0.274	(0.296)
Ta (*4*)	Q	50	0.9861	1.2124 ± 0.0285	0.1498	PA	0.244	0.283	(0.308)
	R	28	0.9867	1.3081 ± 0.0402	0.1199	PA	0.244	0.302	(0.342)
Hm	Q	21	0.9797	1.1677 ± 0.0511	0.1632	PA	(0.240)	0.270	0.289
	R	8	0.9696	1.4434 ± 0.1249	0.0710	PA	(0.197)	0.268	0.321
Kk (*S*)	—	39	0.9793	1.1841 ± 0.0384	0.1451	PA	0.222	0.252	(0.271)
Kk (*B*)	—	19	0.9476	1.4231 ± 0.1043	0.0689	PA	0.183	(0.245)	(0.291)
Kk (*N*)	—	50	0.9151	1.1143 ± 0.0636	0.2538	I	0.330	(0.357)	(0.374)
Sm (*N*)	—	6	0.9584	1.7331 ± 0.2020	0.0470	PA	0.254	(0.423)	(0.569)

In *C. bullatus*, though the sample size and number of samples are small, the ratio of H/L is generally smaller than in *C. vesiculosus* at every fixed size. So far as samples *Hs* (*B*) and *Kk* (*B*) were examined, negatively allometric growth is also apparent.

In *C. nux* only two fossil samples, *Kk* (*N*) and *Sm* (*N*), and one Recent sample, *Bh* (*N*), were morphometrically analyzed, other samples being too small. In sample *Kk* (*N*) the ratio of H/L is evidently smaller than that of *C. vesiculosus* and *C. spinosus* at every growth stage and nearly comparable with that of *C. bullatus*. Negatively allometric growth is also recognized in each valve. On the other hand, the right valves of *Sm* (*N*) show "isometric" growth in this character, though this may be due to the relatively small sample size.

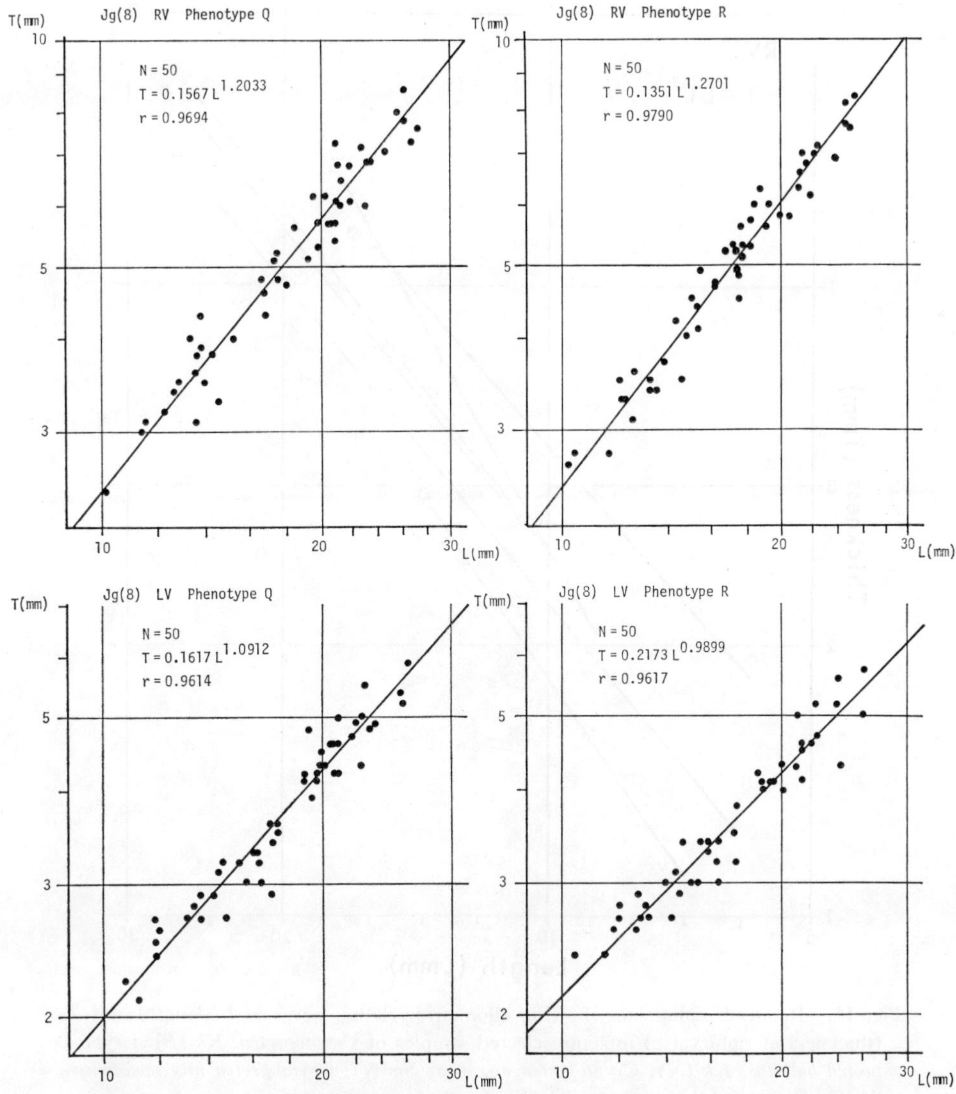

Fig. 14. Example of the allometric relation between L (length) and T (thickness of odd valve). Sample *Jg (8)* of *Cryptopecten vesiculosus*. RV: right valves, LV: left valves. The reduced major axes on double logarithmic scatter diagrams indicate that T is significantly positively allometric to L in right valves but nearly isometric to L in left valves.

3. Length [L] versus Thickness [T]   [Tables 6 and 7]

The shell of *Cryptopecten* is highly inequivalve in every species. Except for the early stages of growth, the right valve is more strongly convex than the left, and the mode of relative growth of the two valves is quite different. The present fossil and dead shell samples are mostly composed of odd valves, and the bivariate analysis on the L-T relation was independently carried out for right and left valves. The curvature of shell

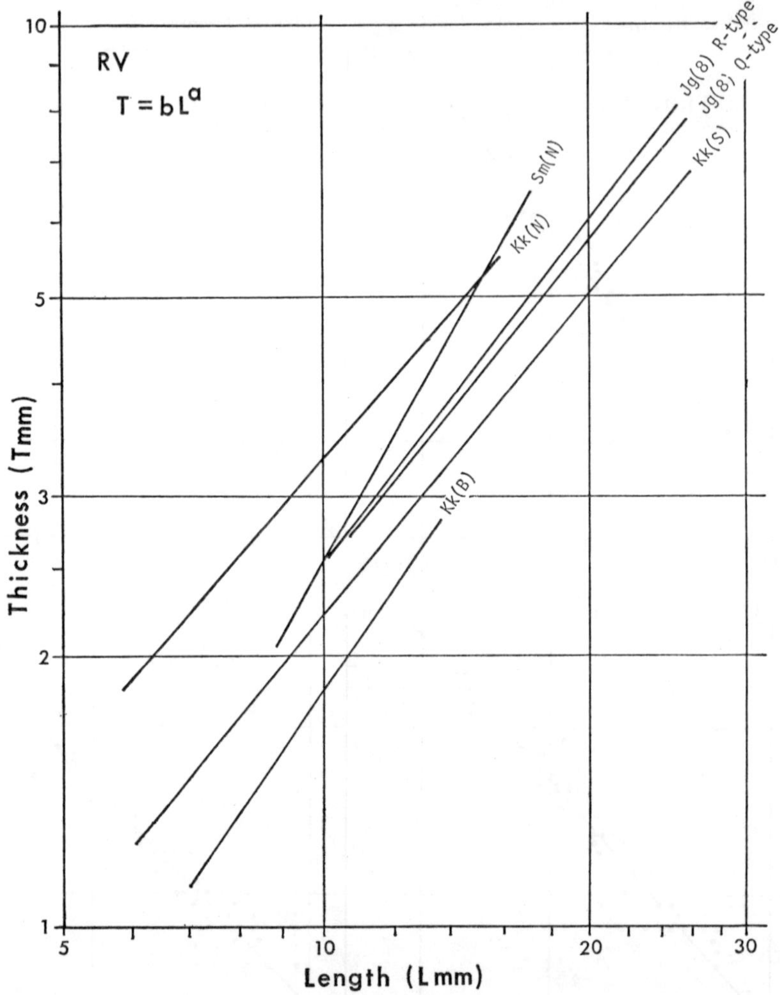

Fig. 15. Reduced major axes showing allometric relation between *L* (length) and *T* (thickness of right valve) in some selected samples of *Cryptopecten*. *Kk* (*B*): *Cryptopecten bullatus*, *Kk* (*N*): *Cryptopecten nux nux*, *Sm(N)*: *Cryptopecten nux sematensis*, *Jg* (*8*): *Cryptopecten vesiculosus*, *Kk* (*S*): *Cryptopecten spinosus*.

surface, as noted before, often changes periodically in accordance with the annual gonad development. Therefore, the shell convexity must be influenced not only by the age composition but also by the season when the individuals died. This is partly responsible, I think, for the relatively low value of correlation coefficient between log *L* and log *T*.

The *L-T* relation for the right valves of *C. vesiculosus* was examined on 18 lots of 11 samples. In all cases the ratio *T/L* becomes much larger with growth, even though minor oscillation may occur during the growth of each individual in relation to periodical change of surface convexity. The correlation coefficient between log *L* and log *T* is generally higher than 0.95, but always decidedly lower than that between log *L* and log *H*. This is evident if the variability in the scatter diagram (Fig. 13) is compared with that of

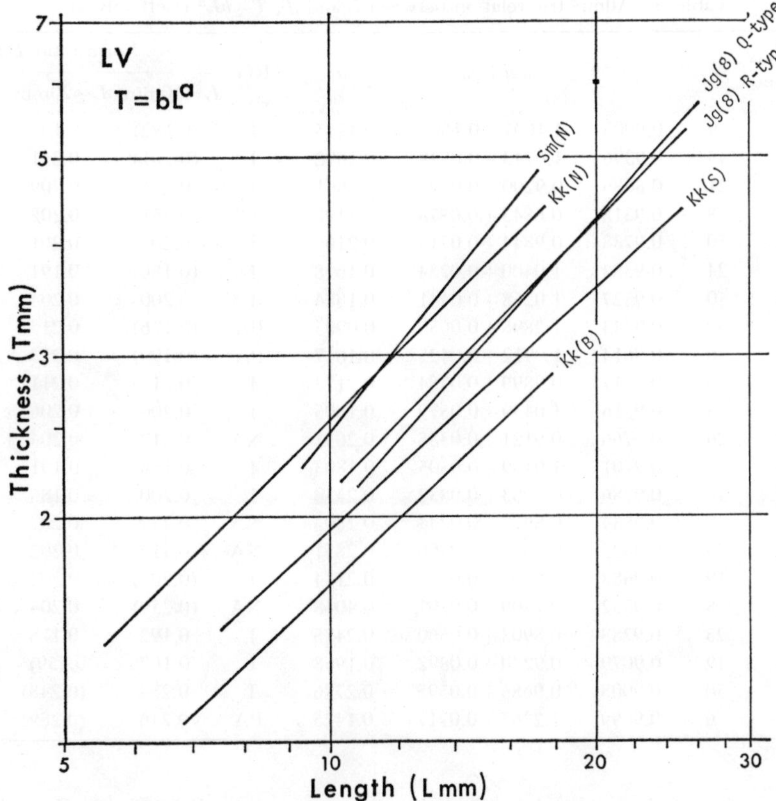

Fig. 16.   Reduced major axes showing allometric relation between *L* (length) and *T* (thickness of left valve) in some selected samples of *Cryptopecten. Kk (B): Cryptopecten bullatus, Kk (N): Cryptopecten nux nux, Sm (N): Cryptopecten nux sematensis, Jg (8): Cryptopecten vesiculosus, Kk (S): Cryptopecten spinosus.*

Fig. 14, which is based on the same material. Except for a few small samples, the growth ratio (*a*) is significantly larger than 1, indicating that *T* is positively allometric against *L*. The difference of *a* is commonly insignificant between the two phenotypes belonging to the same sample, but is sometimes very significant between different samples.

The shell convexity and *L-T* relation in the left valves were also examined on 18 lots of the same samples of *C. vesiculosus*. The ratio of *T/H* is much smaller than that of the right valves after the growth stage of *L*=10 mm. The correlation coefficient (*r*) between log *L* and log *T* is generally lower than the value for the right valves in each sample. Since the ranges of shell size are not much different for the two valves in each sample, this fact seems to indicate that the shell convexity is commonly more variable in the left valves than in the right. The growth ratio (*a*) may or may not exceed 1, and varies considerably with the sample. Of 18 regressions positive allometry and negative allometry were recognized in two and five lots, respectively, and the other eleven lots are regarded as "isometric".

C. spinosus sp. nov., for which only one fossil sample *Kk* (*S*), is available, displays

Table 8 Allometric relation between L and T, $T=bL^a$ (Left valves).

Sample	Pheno-type	N	r	$a\pm\sigma_a$	b	RG	Form ratio T/L		
							$L=10$mm	$L=20$mm	$L=30$mm
Sh 1	Q	9	0.9005	1.1097 ± 0.1609	0.1415	I	(0.182)	0.197	(0.205)
Ik	Q	45	0.9396	1.0743 ± 0.0548	0.1692	I	(0.201)	0.211	0.218
Ob	Q	21	0.8686	0.9200 ± 0.0995	0.2661	I	0.221	0.209	(0.203)
Tm 1	Q	8	0.9318	0.6843 ± 0.0878	0.5198	NA	0.251	0.202	0.178
Ab 1	Q	50	0.9735	0.9844 ± 0.0318	0.2106	I	0.203	0.201	0.200
	R	24	0.9383	1.0400 ± 0.0734	0.1698	I	(0.186)	0.191	0.195
Ms	Q	50	0.9527	1.0268 ± 0.0441	0.1884	I	0.200	0.204	0.206
	R	50	0.9343	1.2898 ± 0.0650	0.0905	PA	(0.176)	0.215	(0.243)
Jg (8)	Q	50	0.9614	1.0912 ± 0.0425	0.1617	PA	(0.199)	0.213	(0.221)
	R	50	0.9617	0.9899 ± 0.0384	0.2173	I	(0.212)	0.211	(0.210)
Su (29)	Q	43	0.9316	1.0439 ± 0.0579	0.1805	I	0.200	0.206	(0.210)
	R	26	0.9766	0.9121 ± 0.0385	0.2651	NA	0.217	0.204	(0.197)
Ec (2)	Q	24	0.9701	1.0189 ± 0.0505	0.1803	I	0.188	0.191	(0.192)
	R	16	0.9286	0.8963 ± 0.0832	0.2538	I	0.200	0.186	(0.178)
Ta (4)	Q	49	0.9834	0.8978 ± 0.0233	0.2805	NA	0.222	0.207	(0.198)
	R	29	0.9825	0.8848 ± 0.0306	0.2851	NA	0.219	0.202	(0.193)
Hm	Q	19	0.9683	0.9801 ± 0.0562	0.2144	I	(0.205)	0.202	(0.200)
	R	8	0.9552	0.7709 ± 0.0807	0.4048	NA	(0.239)	0.204	(0.186)
Kk (S)	—	25	0.9288	0.8903 ± 0.0660	0.2475	I	0.192	0.178	(0.170)
Kk (B)	—	19	0.9079	0.9280 ± 0.0892	0.1968	I	0.167	0.159)	(0.154)
Kk (N)	—	50	0.9008	0.9686 ± 0.0595	0.2726	I	0.254	(0.248)	(0.245)
Sm (N)	—	6	0.9899	1.2365 ± 0.0717	0.1425	PA	0.246	(0.289)	(0.319)

weaker convexity of the two valves, but does not appear to differ from *C. vesiculosus* in mode of relative growth.

C. bullatus also shows much weaker convexity of both valves than *C. vesiculosus*, as is apparent in the ratio T/H at the size of $L=10$ mm. So far as the sample *Kk (B)* was examined, the right valves clearly show positive allometry, but the hypothesis of "isometry" is not rejected in the left valves. In spite of the wide range of shell size in this sample, the correlation coefficient (r) in each valve is considerably low. The variability of this bivariate character is probably somewhat greater in *C. bullatus* than in *C. vesiculosus*.

C. nux, on the other hand, is characterized by a stronger convexity of both valves than other species of *Cryptopecten*. The L-T relation was examined in two fossil samples, *Kk (N)* and *Sm (N)*. Though the sample size of *Sm (N)* is not very large, the mode of relative growth appears to be considerably different between the two samples; the L-T relation in each valve is "isometric" in *Kk (N)* but clearly positively allometric in *Sm (N)*. The ratio of T/L in *Sm (N)* is also much larger than that in *Kk (N)* at the same growth stages.

In all of these four species the right valves always have higher values of growth ratio (a) and lower values of initial growth index (b) than the left valves of the same sample. This indicates a left-convex shell in the very early stages and agrees with the result of SEM observation of the umbonal part of shells. In all the samples the reduced major axes of both valves cross with a large angle, and the shell becomes right-convex, as is indicated by the simple ratios T/L. The allometric relation between L and T in these

Table 9.   Allometric relation between L and D, $D=bL^a$ (Right valves).

Sample	Pheno-type	N	r	$a \pm \sigma_a$	b	RG	Form ratio D/L		
							$L=$10mm	$L=$20mm	$L=$30mm
Ik	Q	23	0.9879	0.7685±0.0249	1.1228	NA	(0.659)	0.561	0.511
Ob	Q	9	0.9096	0.7570±0.1048	1.1685	NA	(0.668)	0.564	0.511
Tm 1	Q	5	0.9258	0.7946±0.1343	1.0929	I	(0.681)	0.591	0.543
Ab 1	Q	50	0.9518	0.7571±0.0328	1.1426	NA	(0.653)	0.552	0.500
	R	42	0.9469	0.8052±0.0399	0.9891	NA	(0.632)	0.552	0.510
Ms	Q	50	0.9527	0.7899±0.0339	1.2072	NA	(0.744)	0.643	0.591
	R	49	0.9452	0.7711±0.0360	1.2981	NA	(0.766)	0.654	0.596
Hy	Q	17	0.9884	0.8055±0.0297	1.1639	NA	0.744	0.650	(0.601)
	R	20	0.9831	0.8227±0.0337	1.1279	NA	0.750	0.663	(0.617)
Jg (8)	Q	50	0.9801	0.7862±0.0221	1.1774	NA	(0.720)	0.620	(0.569)
	R	50	0.9747	0.7737±0.0245	1.2214	NA	(0.725)	0.620	(0.566)
Su (29)	Q	42	0.9613	0.7493±0.0319	1.2852	NA	(0.722)	0.606	(0.548)
	R	31	0.9801	0.7886±0.0281	1.1364	NA	0.698	0.603	(0.554)
Is	Q	50	0.9673	0.6751±0.0242	1.7421	NA	(0.824)	0.658	(0.577)
	R	50	0.9792	0.7525±0.0216	1.3707	NA	(0.775)	0.653	(0.591)
Ec (2)	Q	40	0.9829	0.7944±0.0231	1.1492	NA	0.716	0.621	(0.571)
	R	17	0.9533	0.8657±0.0634	0.9121	NA	0.669	0.610	(0.578)
Ta (4)	Q	50	0.9941	0.8207±0.0126	1.1341	NA	0.751	0.663	(0.616)
	R	27	0.9846	0.7933±0.0267	1.2064	NA	0.750	0.649	(0.597)
Hm	Q	8	0.8658	0.6307±0.1116	1.7653	NA	(0.754)	0.584	0.503
Kk (S)	—	39	0.9869	0.7892±0.0204	1.0466	NA	0.644	0.557	(0.511)
Kk (B)	—	19	0.9840	0.8904±0.0364	0.8760	NA	0.681	(0.631)	(0.603)
Kk (N)	—	50	0.9471	0.8718±0.0396	0.8744	NA	0.651	(0.596)	(0.565)
Sm (N)	—	5	0.9880	0.7937±0.0549	1.2647	NA	0.786	(0.682)	(0.627)
Bh (N)	—	19	0.9811	0.8196±0.0364	1.0868	NA	0.717	(0.633)	(0.588)

species is collectively shown in Figs. 15 and 16 on the basis of several large samples. Similar patterns of L-T relation may be observed in many species of the Pectinidae (expecially Chlamydinae), but such a strong contrast between the valves is probably rare.

4. Length [L] versus Thickness of Conjoined Valves [C]   [Table 8]

Because most samples of living shells are small in size, the relation between length (L) and total thickness (C) was examined on only two samples, Hy and Is, in *C. vesiculosus* and one sample, Bh (N), in *C. nux*. In the former species log L and log C show high correlation, and the growth ratio (a) is significantly larger than 1, indicating the relative growth to be positively allometric. This is probably due to the highly positively allometric growth of the right valve. The values of a are not significantly different for the two phenotypes in each sample. *C. nux* generally shows stronger convexity of two valves than *C. vesiculosus*, but, as is intuitively recognized in Bh (N) and many other samples, the intrapopulational variation of C/L appears to be much greater than that of *C. vesiculosus*. Bivariate analysis on Bh (N) resulted in rather low correlation between log L and log C, and null hypothesis of "isometry" is not rejected. It may be, however, partly due to the small sample size.

Table 10.   Allometric relation between L and D, $D=bL^a$ (Left valves).

Sample	Pheno-type	N	r	$a\pm\sigma_a$	b	RG	Form ratio D/L		
							$L=10$mm	$L=20$mm	$L=30$mm
Ik	Q	21	0.9647	0.8315 ± 0.0478	0.9383	NA	(0.637)	0.566	0.529
$Ab\ 1$	Q	50	0.9799	0.7612 ± 0.0215	1.1392	NA	0.657	0.557	0.506
	R	23	0.9447	0.7879 ± 0.0539	1.0546	NA	(0.647)	0.559	0.513
Ms	Q	49	0.9755	0.8065 ± 0.0253	1.1698	NA	0.749	0.655	0.606
	R	50	0.9510	0.8256 ± 0.0361	1.0866	NA	(0.727)	0.644	(0.600)
Hy	Q	17	0.9884	0.8055 ± 0.0297	1.1639	NA	0.744	0.650	(0.601)
	R	20	0.9831	0.8227 ± 0.0337	1.1279	NA	0.750	0.663	(0.617)
$Jg\ (8)$	Q	50	0.9738	0.7794 ± 0.0251	1.2119	NA	(0.729)	0.626	(0.572)
	R	50	0.9712	0.7779 ± 0.0262	1.1923	NA	(0.715)	0.613	(0.560)
$Su\ (29)$	Q	43	0.9500	0.7614 ± 0.0363	1.2150	NA	0.701	0.594	(0.540)
	R	26	0.9771	0.6913 ± 0.0288	1.4235	NA	0.699	0.565	(0.498)
Is	Q	50	0.9673	0.6751 ± 0.0242	1.7421	NA	(0.824)	0.658	(0.577)
	R	50	0.9792	0.7525 ± 0.0216	1.3707	NA	(0.775)	0.653	(0.591)
$Ec\ (2)$	Q	24	0.9514	0.7967 ± 0.0501	1.1100	NA	0.695	0.604	(0.556)
	R	16	0.9675	0.8551 ± 0.0541	0.9529	NA	0.683	0.617	(0.582)
$Ta\ (4)$	Q	49	0.9794	0.7611 ± 0.0220	1.2672	NA	0.731	0.619	(0.562)
	R	30	0.9906	0.8735 ± 0.0218	0.9792	NA	0.732	0.670	(0.637)
Hm	Q	19	0.9341	0.8069 ± 0.0661	1.0348	NA	(0.663)	0.580	(0.537)
	R	7	0.9941	0.8486 ± 0.0348	0.8964	NA	(0.633)	0.570	(0.536)
$Kk\ (S)$	—	25	0.9872	0.7362 ± 0.0235	1.1972	NA	0.652	0.543	(0.488)
$Kk\ (B)$	—	19	0.9786	0.8920 ± 0.0421	0.8379	NA	0.653	(0.606)	(0.580)
$Kk\ (N)$	—	50	0.9551	0.8367 ± 0.0351	0.9357	NA	0.642	(0.574)	(0.537)
$Sm\ (N)$	—	5	0.9919	0.9550 ± 0.0544	0.7506	I	0.677	(0.656)	(0.644)
$Bh\ (N)$	—	19	0.9811	0.8196 ± 0.0364	1.0868	NA	0.717	(0.633)	(0.588)

5.   Length [L] versus Dorsal Length of Wings [D]   [Tables 9 and 10]

The relation between the overall length (L) and the length of dorsal margin of the two wings (D) was examined on 39 lots of 12 samples of *C. vesiculosus*. These samples are essentially the same as those used in other bivariate analyses, but several individuals were excluded because of broken ends of anterior and/or posterior wings.

The ratio D/L actually seems to indicate the size relation between the disk and the wings. As is indicated by the values of D/L at three fixed sizes, the ratio becomes much smaller with growth. In other words, the auricular part is relatively large in early stages but becomes smaller with growth. This tendency may be ubiquitous in many species of the Pectinidae, but the allometric growth is quite remarkable in this case.

As a result of bivariate analysis, it was found that the correlation coefficient (r) between log L and log D is generally higher than 0.95, and the growth ratio (a) is almost always significantly smaller than 1 with 95 percent confidence, indicating that relative growth is negatively allometric in each valve. Though there are a few exceptions, the values of a are not significantly different for the two phenotypes and nor for the left and right valves in one and the same sample. Fig. 17 is an example of this bivariate analysis.

Meaningful bivariate analysis of L-D relation is now difficult on the Recent samples of *C. bullatus* and *C. nux*, because the sample size is always too small. However, if the individuals belonging to different samples are collectively examined, the auricular part

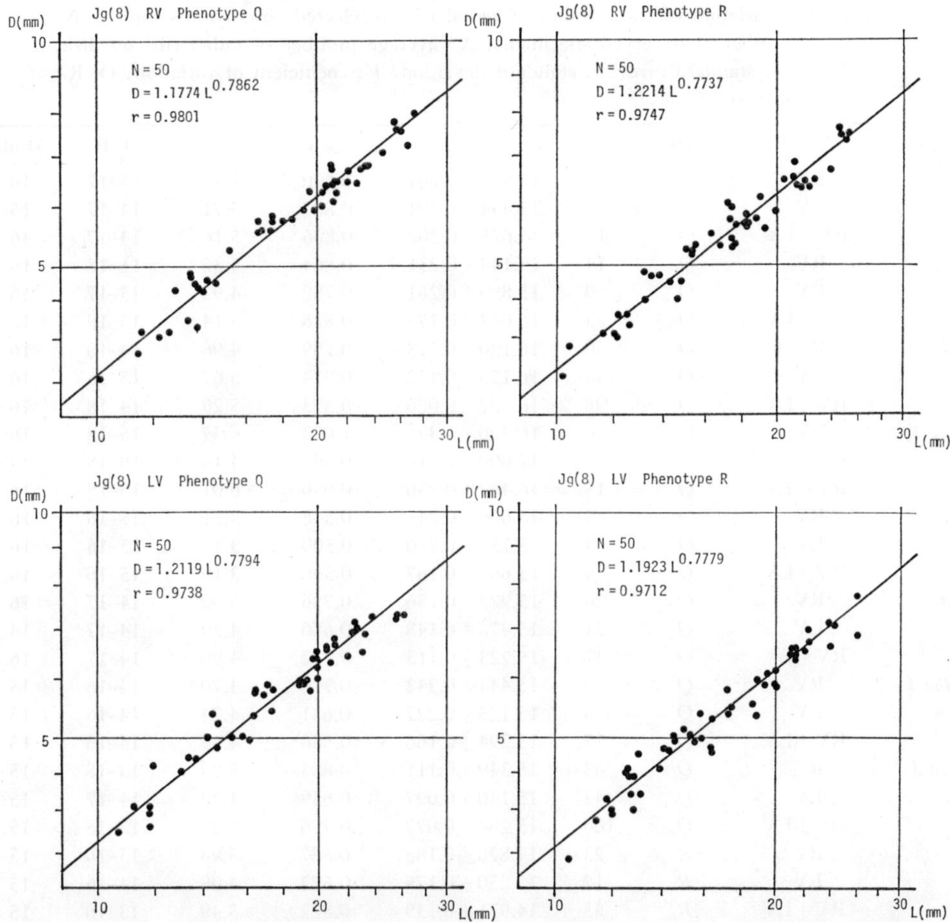

Fig. 17. Example of the allometric relation between L (length) and D (length of dorsal margin). Sample *Jg* (*8*) of *Cryptopecten vesiculosus*. RV: right valves, LV: left valves. The reduced major axes on double logarithmic scatter diagrams indicate that D is significantly negatively allometric to L.

seems to become proportionally smaller in these species as well. Examination of the fossil samples, *Kk* (*S*), *Kk* (*B*) and *Kk* (*N*), showed that the growth ratio (*a*) is significantly smaller than 1 in *C. spinosus*, *C. bullatus* and *C. nux*, indicating evident negative allometry. Therefore, it is concluded that the *L-D* relation is essentially similar in these four species of *Cryptopecten*.

6. Number of Radial Ribs [*X*]   [Table 11]

The number of radial ribs on the disk is a univariate character currently applied for the identification and classification of pectinids. In every species of *Cryptopecten*, radial ribs are quite regular in prominence, neither bifurcated nor, except for very early dissoconch, inserted during the course of growth. Their number would therefore appear to be a growth-invariant character. As noted above (p. 22) however, in many individuals the

48

Table 11. Variation of the number of radial ribs in selected population samples. N: number of observed specimens; $\bar{X}$: average number of radial ribs on disk; $\sigma_{\bar{X}}$: standard error; s: standard deviation; V: coefficient of variation; O. R.: observed range.

Sample	Valve	Type	N	$\bar{X} \pm \sigma_{\bar{X}}$	s	V	O. R.	Mode
Sh 1	RV	Q	7	15.857 ± 0.261	0.690	4.35	15–17	16
	LV	Q	9	15.444 ± 0.294	0.882	5.71	14–17	15
	RV+LV	Q	16	15.625 ± 0.202	0.806	5.16	14–17	16
Nj	RV	Q	14	16.143 ± 0.231	0.864	5.35	15–18	16
	LV	Q	9	15.889 ± 0.261	0.782	4.92	15–17	16
	RV+LV	Q	23	16.043 ± 0.172	0.825	5.14	15–18	16
Ik	RV	Q	50	16.120 ± 0.113	0.799	4.96	14–18	16
	LV	Q	48	16.125 ± 0.132	0.914	5.67	15–18	16
	RV+LV	Q	98	16.122 ± 0.086	0.853	5.29	14–18	16
Kg 1	RV	Q	10	16.200 ± 0.327	1.033	6.37	15–18	16
	LV	Q	5	17.000 ± 0.316	0.707	4.16	16–18	17
	RV+LV	Q	15	16.467 ± 0.256	0.990	6.01	15–18	16
Mz 2	RV	Q	5	15.600 ± 0.245	0.548	3.51	15–16	16
	LV	Q	4	15.750 ± 0.250	0.500	3.17	15–16	16
	RV+LV	Q	9	15.667 ± 0.167	0.500	3.19	15–16	16
Ob	RV	Q	26	15.923 ± 0.156	0.796	5.00	14–17	16
	LV	Q	21	15.476 ± 0.148	0.680	4.39	14–17	16
	RV±LV	Q	47	15.723 ± 0.113	0.772	4.90	14–17	16
Tm 1+2	RV	Q	9	15.444 ± 0.242	0.726	4.70	14–16	15
	LV	Q	8	15.125 ± 0.227	0.641	4.24	14–16	15
	RV+LV	Q	17	15.294 ± 0.166	0.686	4.49	14–16	15
Sn 3	RV	Q	63	15.349 ± 0.111	0.883	5.75	13–18	15
	LV	Q	43	15.140 ± 0.097	0.639	4.22	14–17	15
	RV+LV	Q	106	15.264 ± 0.077	0.796	5.22	13–18	15
	RV	R	23	14.826 ± 0.185	0.887	5.98	13–16	15
	LV	R	12	15.250 ± 0.179	0.622	4.08	14–16	15
	RV+LV	R	35	14.971 ± 0.139	0.822	5.49	13–16	15
	RV+LV	$Q+R$	141	15.191 ± 0.068	0.810	5.33	13–18	15
Ab 1	RV	Q	50	15.120 ± 0.109	0.773	5.11	14–16	15
	LV	Q	50	14.920 ± 0.081	0.570	3.81	14–17	15
	RV+LV	Q	100	15.040 ± 0.068	0.680	4.52	14–17	15
	RV	R	42	14.857 ± 0.126	0.814	5.48	13–17	15
	LV	R	24	14.917 ± 0.146	0.717	4.81	14–17	15
	RV+LV	R	66	14.879 ± 0.095	0.775	5.21	13–17	15
	RV+LV	$Q+R$	166	14.976 ± 0.056	0.722	4.82	13–17	15
Jz	RV	Q	50	15.080 ± 0.137	0.966	6.40	13–18	15
	LV	Q	50	15.060 ± 0.101	0.712	4.73	14–17	15
	RV+LV	Q	100	15.070 ± 0.084	0.844	5.60	13–18	15
	RV	R	50	14.840 ± 0.108	0.766	5.16	13–17	15
	LV	R	50	14.880 ± 0.109	0.773	5.19	13–17	15
	RV+LV	R	100	14.860 ± 0.077	0.766	5.15	13–17	15
	RV+LV	$Q+R$	200	14.965 ± 0.057	0.811	5.42	13–18	15
Sm	RV	Q	50	14.780 ± 0.112	0.790	5.34	13–17	15
	LV	Q	50	14.580 ± 0.095	0.673	4.61	13–16	15
	RV+LV	Q	100	14.680 ± 0.074	0.737	5.02	13–17	15
	RV	R	13	14.538 ± 0.215	0.776	5.34	13–16	15
	LV	R	12	14.417 ± 0.193	0.669	4.64	13–15	14

Table 11   (cont'd—2)

Sample	Valve	Type	N	$\bar{X}\pm\sigma_{\bar{X}}$	s	V	O. R.	Mode
	RV+LV	R	25	14.480±0.143	0.714	4.93	13–16	14
	RV+LV	Q+R	125	14.640±0.066	0.734	5.01	13–17	15
Ms	RV	Q	50	15.160±0.117	0.828	5.46	14–17	15
	LV	Q	50	15.020±0.117	0.829	5.52	13–17	15
	RV+LV	Q	100	15.090±0.084	0.842	5.58	13–17	15
	RV	R	50	15.140±0.113	0.797	5.26	14–17	15
	LV	R	50	15.060±0.090	0.638	4.24	14–17	15
	RV+LV	R	100	15.100±0.073	0.732	4.85	14–17	15
	RV+LV	Q+R	200	15.095±0.056	0.787	5.21	13–17	15
Hy	RV	Q	18	14.667±0.198	0.840	5.73	13–16	15
	LV	Q	18	14.556±0.202	0.856	5.88	13–16	15
	RV+LV	Q	36	14.611±0.140	0.838	5.73	13–16	15
	RV	R	20	14.700±0.147	0.657	4.47	14–16	15
	LV	R	20	14.950±0.135	0.605	4.05	14–16	15
	RV+LV	R	40	14.825±0.101	0.636	4.29	14–16	15
	RV+LV	Q+R	76	14.724±0.085	0.741	5.03	13–16	15
Jg (8)	RV	Q	50	14.720±0.114	0.809	5.50	13–17	15
	LV	Q	50	14.520±0.100	0.707	4.87	13–16	14
	RV+LV	Q	100	14.620±0.076	0.763	5.22	13–17	15
	RV	R	50	14.500±0.104	0.735	5.07	13–16	14
	LV	R	50	14.600±0.111	0.782	5.36	13–16	15
	RV+LV	R	100	14.550±0.076	0.757	5.20	13–16	15
	RV+LV	Q+R	200	14.585±0.054	0.759	5.20	13–17	15
Ts (11)	RV	Q	20	15.050±0.153	0.686	4.56	14–16	15
	LV	Q	24	15.042±0.141	0.690	4.59	14–17	15
	RV+LV	Q	44	15.045±0.103	0.680	4.52	14–17	15
	RV	R	15	14.867±0.236	0.915	5.16	13–16	15
	LV	R	15	14.867±0.165	0.640	4.30	14–16	15
	RV+LV	R	30	14.867±0.142	0.776	5.22	13–16	15
	RV+LV	Q+R	74	14.973±0.084	0.721	4.82	13–17	15
Su (29)	RV	Q	44	14.750±0.113	0.751	5.09	13–16	15
	LV	Q	43	14.605±0.129	0.849	5.81	13–17	15
	RV+LV	Q	87	14.678±0.086	0.800	5.45	13–17	15
	RV	R	32	14.594±0.126	0.712	4.88	13–16	15
	LV	R	26	14.538±0.138	0.706	4.86	13–16	15
	RV+LV	R	58	14.569±0.092	0.704	4.83	13–16	15
	RV+LV	Q+R	145	14.634±0.063	0.762	5.21	13–17	15
Is	RV	Q	50	14.700±0.119	0.839	5.71	13–16	15
	LV	Q	50	14.620±0.117	0.830	5.68	13–16	15
	RV+LV	Q	100	14.660±0.083	0.831	5.67	13–16	15
	RV	R	50	14.420±0.095	0.673	4.67	13–16	14
	LV	R	50	14.500±0.108	0.763	5.26	13–16	14, 15
	RV+LV	R	100	14.460±0.072	0.717	4.96	13–16	14
	RV+LV	Q+R	200	14.560±0.055	0.781	5.36	13–16	14, 15
Km (1)	RV	Q	28	14.750±0.151	0.799	5.42	13–16	15
	LV	Q	30	14.800±0.147	0.805	5.44	13–17	15
	RV+LV	Q	58	14.776±0.104	0.796	5.38	13–17	15
	RV	R	13	14.769±0.257	0.927	6.28	13–16	15
	LV	R	18	14.778±0.152	0.647	4.38	14–16	15
	RV+LV	R	31	14.774±0.137	0.762	5.16	13–16	15

Table 11   (cont'd—3)

Sample	Valve	Type	N	$\bar{X}\pm\sigma_{\bar{X}}$	s	V	O. R.	Mode
	RV+LV	$Q+R$	89	14.775 ± 0.083	0.780	5.28	13–17	15
Bt	RV	Q	22	15.045 ± 0.139	0.653	4.34	14–16	15
	LV	Q	16	15.063 ± 0.143	0.574	3.81	14–16	15
	RV+LV	Q	38	15.053 ± 0.099	0.613	4.07	14–16	15
	RV	R	11	14.909 ± 0.163	0.539	3.62	14–16	15
	LV	R	5	15.000 ± 0.316	0.707	4.71	14–16	15
	RV+LV	R	16	14.938 ± 0.143	0.574	3.84	14–16	15
	RV+LV	$Q+R$	54	15.019 ± 0.081	0.598	3.98	14–16	15
Ec (2)	RV	Q	40	15.250 ± 0.117	0.742	4.87	14–17	15
	LV	Q	24	15.125 ± 0.139	0.680	4.49	14–16	15
	RV+LV	Q	64	15.203 ± 0.090	0.717	4.71	14–17	15
	RV	R	17	15.118 ± 0.169	0.697	4.61	14–16	15
	LV	R	16	15.250 ± 0.194	0.775	5.08	14–17	15
	RV+LV	R	33	15.182 ± 0.127	0.727	4.79	14–17	15
	RV+LV	$Q+R$	97	15.196 ± 0.073	0.716	4.71	14–17	15
Ta (4)	RV	Q	50	15.120 ± 0.120	0.849	5.61	13–17	15
	LV	Q	50	15.200 ± 0.103	0.728	4.79	14–17	15
	RV+LV	Q	100	15.160 ± 0.079	0.788	5.20	13–17	15
	RV	R	28	15.036 ± 0.120	0.637	4.24	14–16	15
	LV	R	30	15.167 ± 0.136	0.747	4.92	14–17	15
	RV+LV	R	58	15.103 ± 0.091	0.693	4.59	14–17	15
	RV+LV	$Q+R$	158	15.139 ± 0.060	0.753	4.97	13–17	15
Hm	RV	Q	21	15.429 ± 0.148	0.676	4.38	14–16	15
	LV	Q	19	15.263 ± 0.150	0.653	4.28	14–17	15
	RV+LV	Q	40	15.350 ± 0.105	0.662	4.31	14–17	15
	RV	R	8	15.000 ± 0.267	0.756	5.04	14–16	15
	LV	R	8	14.875 ± 0.227	0.641	4.31	14–16	15
	RV+LV	R	16	14.938 ± 0.170	0.680	4.55	14–16	15
	RV+LV	$Q+R$	56	15.232 ± 0.092	0.687	4.51	14–17	15
Kk (S)	RV	—	39	13.949 ± 0.116	0.724	5.19	12–15	14
	LV	—	25	13.800 ± 0.115	0.577	4.18	13–15	14
	RV+LV	—	64	13.891 ± 0.084	0.669	4.82	12–15	14
Hs (B)	RV	—	7	19.897 ± 0.261	0.690	3.48	19–21	20
	LV	—	8	20.000 ± 0.327	0.926	4.63	19–22	20
	RV+LV	—	15	19.933 ± 0.206	0.799	4.01	19–22	20
Kk (B)	RV	—	19	21.263 ± 0.185	0.806	3.79	20–23	21
	LV	—	19	21.263 ± 0.200	0.872	4.10	20–23	21
	RV+LV	—	38	21.263 ± 0.134	0.828	3.89	20–23	21
Kk (N)	RV	—	50	19.580 ± 0.125	0.883	4.51	18–22	20
	LV	—	50	19.580 ± 0.140	0.992	5.06	18–22	20
	RV+LV	—	100	19.580 ± 0.093	0.934	4.77	18–22	20
Bh (N)	RV	—	19	20.842 ± 0.175	0.765	3.67	19–22	21
	LV	—	19	20.789 ± 0.181	0.787	3.79	19–22	21
Sm (N)	RV	—	6	19.167 ± 0.307	0.753	3.93	18–20	19
	LV	—	6	19.000 ± 0.365	0.894	4.71	18–20	19
	RV+LV	—	12	19.083 ± 0.229	0.793	4.16	18–20	19
Mn (Y)	RV	—	12	21.500 ± 0.314	1.087	5.06	20–24	22
	LV	—	9	21.444 ± 0.377	1.130	5.27	20–23	21
	RV+LV	—	21	21.476 ± 0.235	1.078	5.02	20–24	21

outermost radial ribs near the anterodorsal and posterodorsal margins of disk are much weakened or have vanished altogether in the course of growth. I thus define X to be the number of radial ribs counted at the middle stage of growth (approximately 10 mm in height in *C. vesiculosus*, *C. bullatus* and *C. yanagawaensis*; approximately 7 mm in *C. spinosus* and *C. nux*).

This character was examined on 135 lots of 22 samples of *C. vesiculosus*, three lots of one sample of *C. spinosus*, six lots of two samples of *C. bullatus*, six lots of two samples of *C. nux*, and three lots of one sample of *C. yanagawaensis*. The results are collectively given in Table 11, where the mean ($\bar{X}$), standard error ($\sigma_{\bar{X}}$), standard deviation (s), coefficient of variation (V), observed range (O. R) and mode of frequency distribution ($\hat{X}$) are indicated.

In *C. vesiculosus* the mode commonly lies at 14 or 15 in Recent and younger fossil samples, but at 16 in such older fossil samples such as *Ik* and *Kg 1*. The histograms are almost always unimodal and resemble normal frequency distributions. The uniformity of variability of this character is worthy of notice; the standard deviation (s) is almost always confined to the range from 3.0 to 6.0. T-tests show that the values of $\bar{X}$ for each of the two valves in the same sample are seldom significantly different. The same is probably true for the two phenotypes, though in many cases Phenotype R seems to show a slightly smaller value of $\bar{X}$ than Phenotype Q. This is probably because the radial ribs of Phenotype R are weaker and less conspicuous, especially near the peripheries of disk, than those of Phenotype Q.

One of the distinguishing characters of *C. spinosus* is, as shown in the sample Kk (S), the significantly smaller number of radial ribs in comparison with the other species of *Cryptopecten*; the number ranges from 12 to 15, and the mode lies at 14.

C. bullatus has, meanwhile, more numerous radial ribs than *C. vesiculosus*. Though every Recent sample consists of a small number of individuals, the values of $\bar{X}$ are approximately 20 in the samples USNM 764152 and *Hs* (B). This is not much different from the value of the Hawaiian and Philippine specimens including the holotype of *C. alli*. The fossil sample *Kk* (B), which is identical with *C. bullatus* in various other characters, nevertheless shows a somewhat larger number of radial ribs.

C. nux also possesses more numerous radial ribs than *C. vesiculosus*. Most specimens of the fossil samples *Kk* (N) and *Sm* (N) as well as several Recent and fossil specimens from various localities have 18 to 22 radial ribs.

C. yanagawaensis seems to have the most numerous radial ribs of the five species of *Cryptopecten*. In the solitary fossil sample *Mn* (Y) most specimens possess 20 to 25 radial ribs with the mode lying at 21 or 22. The variability of this character may not be so great as was suggested by Masuda (1958).

Although it is unknown whether this character is entirely controlled by genetic factors or not, I conclude that the number of radial ribs of *Cryptopecten*, is almost entirely growth-invariant and can be regarded as a useful univariate meristic character for discussions of geographic variation and evolutionary change as well as specific distinction.

B. Discontinuous Variation

Discontinuous variation is a phenomenon which is characterized by the presence of

52

two or more discrete phenotypes within an interbreeding population. It may or may not
be controlled by genetic factors. The term "polymorphism" is sometimes used for any
kind of discontinuous variation, but, as was discussed by Mayr (1963, p. 150), such
nongenetic discontinuous variations as age, social, ecological and traumatic variations are
preferably distinguished from true polymorphism and treated under a different term,
polyphenism. Ford (1940, 1964) defined genetic polymorphism as "the occurrence together in the same locality of two or more discontinuous forms of a species in such pro-
portion that the rarest of them cannot be maintained by recurrent mutation".

Most genetic approaches to evolution at the morphological level have been carried out
on the basis of polymorphism, simply because the gene frequency of populations and its
change in time and space can be easily recognized phenotypically. In marine organisms
the genetic background of discontinuous variation is often difficult to determine. Still,
discrimination between polymorphism and polyphenism is possible through careful
observations of morphology, population structure and ecological habit as well as through
analogy with well-experimented organisms.

1. Color Polymorphism

Polymorphism of shell coloration seems to be widespread in molluscs. In certain
groups of the Bivalvia, ground color and color pattern vary discontinuously within one
and the same population. Discontinuous variation of ground color is also quite striking
in several species of *Chlamys*, *Aequipecten*, and some other genera of the Pectinidae. In
the beautiful Japanese species, *Chlamys* (*Mimachlamys*) *nobilis* (Reeve), for example,
most individuals are reddish brown in color, but orange, yellow, purple, and white
individuals are occasionally found together in the same population. So far as I am aware,
no Mendelian experiment has been successfully carried out, but the color variation of
pectinids is probably genetic. In some land snails polymorphism in ground coloration
and color banding is controlled by a series of multiple alleles constituting a supergene
(Cain and Sheppard, 1950; Komai and Emura, 1955; etc.).

In the four extant species of *Cryptopecten*, most individuals are reddish-brown to
magenta in ground color, frequently mottled with irregular oblique bands or spots of pale
coloration. Certain individuals (less than 5 percent in *C. vesiculosus*) are yellow or orange
in ground color. Though yellow and orange individuals are sometimes not clearly separable from each other, they are never mottled but quite monotonic in coloration. The
phenotypic frequency of yellow plus orange individuals can thus be clearly recognized
in Recent samples, and it would be useful for the analysis of geographic variation if very
large fresh samples could be examined. These features of color polymorphism are essentially similar in *C. vesiculosus*, *C. bullatus*, *C. nux* and *C. phrygium*, suggesting that
these species are closely related to one another and have a common ancestry.

Color polymorphism of *Cryptopecten* is an interesting phenomenon but generally
difficult to observe in fossil samples. Mottled color pattern of Late Pleistocene fossil
specimens are often visible (generally more clearly under ultraviolet light), but variation
of ground color is hard to recognize.

2. Dimorphism of Surface Sculpture in *C. vesiculosus* [Table 12]

As pointed out previously (Hayami, 1972, 1973), every sample of *C. vesiculosus* after

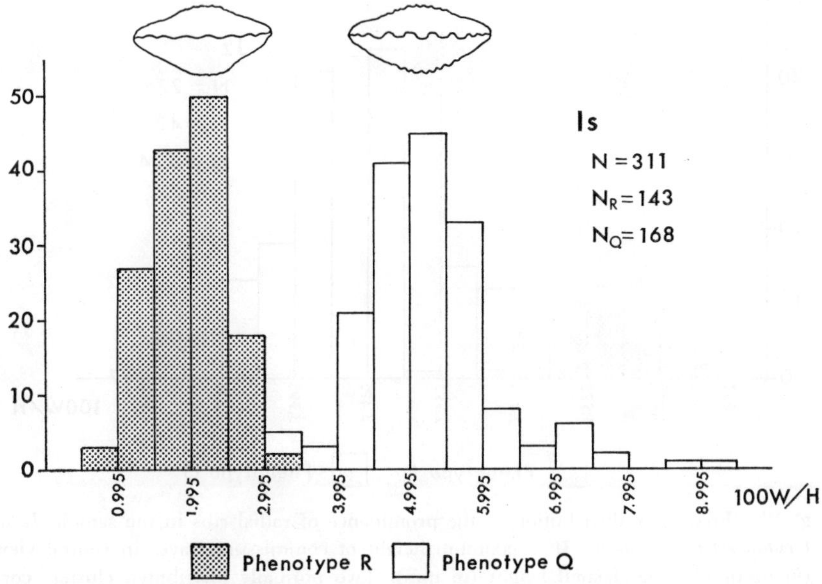

Fig. 18.   Frequency distribution of the prominence of radial ribs in the sample *Is*
of *Cryptopecten vesiculosus*. *W*: maximum height of commissure waves in ventral view
(in mm), *H*: overall shell height (in mm). The prominence of radial ribs is roughly
expressed by the ratio 100 *W*/*H*. Two normally distributed clusters well correspond
with the two phenotypes. Adapted from Hayami (1973, text-fig. 4).

the Middle Pleistocene consists of two discrete phenotypes, called Q and R, without any
intermediate individual. The morphological difference between these becomes clear after
the beginning of the third stage of dissoconch (see p. 23). The fundamental radial rib
structure, consisting of a central solid ridge and lateral hollow parts, is the same for the
two phenotypes, but in transverse section the radial ribs of Phenotype Q are clearly more
quadrate and highly elevated than those of Phenotype R. This difference is well reflected
in the ratio between the overall height (H) and the maximum wave height of the commissure (W) in ventral view as studied previously on the samples *Is* and *Jz* (in part)
(Hayami, 1973, p. 406, figs. 4 and 5: here reproduced in Figs. 18 and 19). Clearly bimodal
histograms of the ratio W/H were obtained and the two clusters correspond well with
the two phenotypes.

A more conspicuous difference between the two phenotypes exists in the sculpture on
the interspaces of radial ribs, especially in the inclination of scales and the shape of
hollow parts, as will be described in detail in the systematic description (p. 23). The
scales covering the hollow parts in both phenotypes are frequently exfoliated in abraded
dead shells and fossils. Nevertheless, the distinction between the two phenotypes is
always quite clear and never arbitrary. Therefore the phenotypic frequency can be precisely determined in every sample. The results are collectively shown in Table 12.

The individuals belonging to Phenotype Q seem to be further separable into two sub-
phenotypes by the presence or absence of erect scales on the top of the central ridges (see
Plate 4, Figs. 1, 2, Plate 5, Figs. 1, 2). The two subphenotypes, called *rough* type and

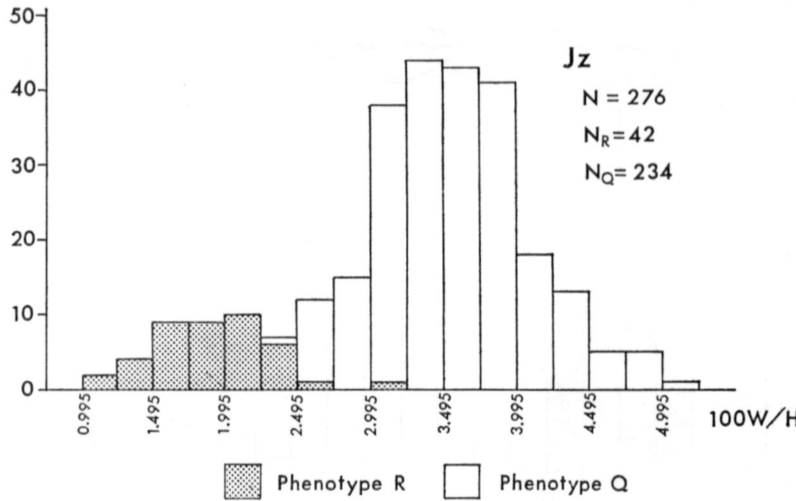

Fig. 19. Frequency distribution of the prominence of radial ribs in the sample *Jz* of *Cryptopecten vesiculosus*. *W*: maximum height of commissure waves in ventral view (in mm), *H*: overall shell height (in mm). Two normally distributed clusters correspond with the two phenotypes. The difference of mean values in each phenotype from those of the sample *Is* may be partly due to the unfavorable condition of this material, in which the ventral margin of shell (especially the projected portion of ribs) is often worn or slightly broken. Adapted from Hayami (1973, text-fig. 5).

smooth type, are discontinuous with few intermediate individuals. For example, among 169 individuals belonging to Phenotype *Q* of the sample *Is*, 60 and 109 individuals are assigned to the *rough* type and *smooth* type, respectively. The frequency of the two sub-phenotypes is also recognizable, though somewhat arbitrarily, in some well-preserved dead shell and fossil samples (Plate 6, Figs. 1, 2), but their discrimination is generally difficult in the case of abraded specimens and individuals smaller than 10 mm in length. The top of radial ribs in Phenotype *R* is, on the contrary, always quite smooth, lacking any erect scales. This fact may suggest that the discontinuous variation between the two subphenotypes is controlled by some cause related to the factor determining the phenotypic difference. Although the *smooth* subphenotype appears at a glance to be morphologically intermediate between the *rough* subphenotype and Phenotype *R*, the hypothesis of Hardy-Weinberg equilibrium, that the first were heterozygote individuals between two homozygote ones, is statistically rejected in *Is* and some other large samples.

Judging from the following facts, these phenotypes are taxonomically inseparable in spite of the clear morphological discontinuity.

1) Living individuals belonging to the two phenotypes were always found together within the same dredge sample (or subsample) (e. g., *Kz*, *Jg* (*1–9*, *16–23*), *Zs* (*4, 8*), *Hk*) (Table 1).

2) The two phenotypes are coexistent in all the large samples of Recent dead shells as well as fossils after Middle Pleistocene. The phenotypic frequency is considerably stable in the same sea area, though some geographic variation and chronological change

Table 12. Phenotypic frequency in each sample of *Cryptopecten vesiculosus*. N: total number of individuals; N_R: number of individuals belonging to Phenotype R; N_Q: number of individuals belonging to Phenotype Q.

Sample	Valve	N	N_R	N_Q	N_R/N	95% range	Age
Sh 1	RV	101	0	101	0.000	0.000–0.037	ca. 3.5 Ma
	LV	128	0	128	0.000	0.000–0.029	
	Total	229	0	229	0.000	0.000–0.016	
Sh 2	RV	7	0	7	0.000	0.000–0.354	ca. 3.5 Ma
	LV	9	0	9	0.000	0.000–0.299	
	Total	16	0	16	0.000	0.000–0.194	
Ht	RV	5	0	5	0.000	0.000–0.434	ca. 3.0 Ma
	LV	2	0	2	0.000	0.000–0.658	
	Total	7	0	7	0.000	0.000–0.354	
Iy	RV	11	0	11	0.000	0.000–0.259	2.5–2.0 Ma
	LV	10	0	10	0.000	0.000–0.278	
	Total	21	0	21	0.000	0.000–0.155	
Nj	RV	16	0	16	0.000	0.000–0.194	unknown
	LV	13	0	13	0.000	0.000–0.228	
	Total	29	0	29	0.000	0.000–0.155	
Ik	CV	8	0	8	0.000	0.000–0.324	2.5–2.0 Ma
	RV	73	0	73	0.000	0.000–0.050	
	LV	73	0	73	0.000	0.000–0.050	
	Total	154	0	154	0.000	0.000–0.024	
Sw	RV	1	0	1	0.000	0.000–0.793	unknown
	LV	2	0	2	0.000	0.000–0.658	
	Total	3	0	3	0.000	0.000–0.561	
Ne	RV	14	0	14	0.000	0.000–0.215	unknown
	LV	18	0	18	0.000	0.000–0.176	
	Total	32	0	32	0.000	0.000–0.107	
Kg 1	RV	6	0	6	0.000	0.000–0.390	2.4–1.9 Ma
	LV	14	0	14	0.000	0.000–0.215	
	Total	20	0	20	0.000	0.000–0.161	
Kg 2	RV	4	0	4	0.000	0.000–0.490	2.4–1.9 Ma
	LV	2	0	2	0.000	0.000–0.658	
	Total	6	0	6	0.000	0.000–0.390	
Mz 1	RV	2	0	2	0.000	0.000–0.658	unknown
	LV	3	0	3	0.000	0.000–0.561	
	Total	5	0	5	0.000	0.000–0.434	
Mz 2	RV	7	0	7	0.000	0.000–0.354	unknown
	LV	6	0	6	0.000	0.000–0.390	
	Total	13	0	13	0.000	0.000–0.228	
Ob	RV	46	0	46	0.000	0.000–0.077	2.0–1.7 Ma
	LV	39	0	39	0.000	0.000–0.090	
	Total	85	0	85	0.000	0.000–0.043	
Oe	RV	1	0	1	0.000	0.000–0.793	unknown
	LV	1	0	1	0.000	0.000–0.793	
	Total	2	0	2	0.000	0.000–0.658	
Tm 1	RV	20	0	20	0.000	0.000–0.161	1.0–0.7 Ma
	LV	16	0	16	0.000	0.000–0.194	
	Total	36	0	36	0.000	0.000–0.096	
Tm 2	RV	6	0	6	0.000	0.000–0.390	1.0–0.7 Ma
	LV	5	0	5	0.000	0.000–0.434	

Table 12. (cont'd—2)

Sample	Valve	N	N_R	N_Q	N_R/N	95% range	Age
	Total	11	0	11	0.000	0.000–0.259	
Ij 1	RV	12	0	12	0.000	0.000–0.242	ca. 0.6 Ma
	LV	13	0	13	0.000	0.000–0.228	
	Total	25	0	25	0.000	0.000–0.133	
Ij 2	RV	4	0	4	0.000	0.000–0.490	ca. 0.6 Ma
	LV	4	0	4	0.000	0.000–0.490	
	Total	8	0	8	0.000	0.000–0.324	
Mt 1	RV	4	0	4	0.000	0.000–0.490	unknown
	LV	6	0	6	0.000	0.000–0.390	
	Total	10	0	10	0.000	0.000–0.278	
Mt 2	RV	2	0	2	0.000	0.000–0.658	unknown
	LV	3	0	3	0.000	0.000–0.561	
	Total	5	0	5	0.000	0.000–0.434	
Nk	RV	1	0	1	0.000	0.000–0.793	unknown
Sn 1	RV	2	0	2	0.000	0.000–0.658	ca. 0.5 Ma
	LV	3	2	1	0.667	0.208–0.939	
	Total	5	2	3	0.400	0.118–0.769	
Sn 2	RV	17	3	14	0.176	0.062–0.410	ca. 0.5 Ma
	LV	11	0	11	0.000	0.000–0.259	
	Total	28	3	25	0.107	0.037–0.272	
Sn 3	RV	110	28	82	0.255	0.183–0.344	ca. 0.5 Ma
	LV	67	13	54	0.194	0.117–0.304	
	Total	177	41	136	0.232	0.176–0.299	
Si	RV	1	0	1	0.000	0.000–0.793	ca. 0.45 Ma
Nn	LV	1	0	1	0.000	0.000–0.793	unknown
Hg	RV	47	8	39	0.170	0.089–0.301	ca. 0.37 Ma
	LV	8	1	7	0.125	0.022–0.471	
	Total	55	9	46	0.164	0.089–0.283	
Ny 1	RV	63	8	55	0.127	0.066–0.231	ca. 0.37 qa
	LV	31	4	27	0.129	0.051–0.288	
	Total	94	12	82	0.128	0.075–0.210	
Ny 2	RV	35	9	26	0.257	0.142–0.421	ca. 0.37 Ma
	LV	20	2	18	0.100	0.028–0.301	
	Total	55	11	44	0.200	0.116–0.324	
Ny 3	RV	13	4	9	0.308	0.127–0.577	ca. 0.37 Ma
	LV	7	1	6	0.143	0.026–0.513	
	Total	20	5	15	0.250	0.112–0.469	
Ny 4	RV	15	3	12	0.200	0.070–0.452	ca. 0.37 Ma
	LV	47	7	40	0.149	0.074–0.277	
	Total	62	10	52	0.161	0.090–0.272	
Jz	RV	819	104	715	0.127	0.106–0.152	ca. 0.37 Ma
	LV	912	114	798	0.125	0.105–0.148	
	Total	1,731	218	1,513	0.126	0.111–0.142	
Ic	RV	46	5	41	0.109	0.048–0.231	ca. 0.37 Ma
	LV	61	11	50	0.180	0.104–0.294	
	Total	107	16	91	0.150	0.095–0.230	
Ab 1	RV	326	63	263	0.193	0.154–0.239	ca. 0.37 Ma
	LV	387	66	321	0.171	0.137–0.212	
	Total	713	129	584	0.181	0.154–0.211	
Ab 2	CV	2	1	1	0.500	0.095–0.905	ca. 0.37 Ma

Table 12. (cont'd—3)

Sample	Valve	N	N_R	N_Q	N_R/N	95% range	Age
	RV	122	16	106	0.131	0.071–0.229	
	LV	191	26	165	0.136	0.095–0.192	
	Total	315	43	272	0.137	0.103–0.179	
Nm	RV	16	4	12	0.250	0.102–0.495	unknown
	LV	8	2	6	0.250	0.071–0.591	
	Total	24	6	18	0.250	0.120–0.449	
Sm	RV	126	20	106	0.159	0.105–0.233	ca. 0.29 Ma
	LV	124	19	105	0.153	0.100–0.227	
	Total	250	39	211	0.156	0.116–0.206	
Tn	RV	2	0	2	0.000	0.000–0.658	ca. 0.15 Ma
Ys	RV	8	3	5	0.375	0.137–0.694	unknown
	LV	6	2	4	0.333	0.097–0.700	
	Total	14	5	9	0.357	0.163–0.612	
Ms	RV	716	162	554	0.226	0.197–0.258	ca. 0.003 Ma
	LV	694	138	556	0.199	0.171–0.230	
	Total	1,410	300	1,110	0.213	0.192–0.235	
Tt	RV	11	3	8	0.273	0.098–0.566	Recent
	LV	12	5	7	0.417	0.194–0.681	
	Total	23	8	15	0.348	0.188–0.551	
Kz	CV	2	1	1	0.500	0.095–0.905	Recent
Hy	CV	38	20	18	0.526	0.372–0.675	Recent
	RV	124	56	68	0.452	0.367–0.540	
	LV	190	83	107	0.437	0.368–0.508	
	Total	352	159	193	0.452	0.401–0.504	
Jg (1–26)	CV	267	129	138	0.483	0.424–0.543	Recent
	RV	5,370	2,235	3,135	0.416	0.403–0.429	
	LV	5,136	2,199	2,937	0.428	0.415–0.442	
	Total	10,773	4,563	6,210	0.424	0.414–0.433	
Ma	CV	1	1	0	1.000	0.207–1.000	Recent
	RV	43	14	29	0.325	0.205–0.475	
	LV	26	11	15	0.423	0.255–0.611	
	Total	70	26	44	0.371	0.268–0.488	
Am	RV	125	50	75	0.400	0.318–0.488	Recent
	LV	98	37	61	0.378	0.288–0.477	
	Total	223	87	136	0.390	0.310–0.476	
Aj (1–2)	CV	3	1	2	0.333	0.061–0.792	Recent
Os	CV	2	2	0	1.000	0.342–1.000	Recent
Zs (1–9)	CV	9	5	4	0.556	0.267–0.812	Recent
	RV	59	24	35	0.407	0.291–0.534	
	LV	33	11	22	0.333	0.197–0.504	
	Total	101	40	61	0.396	0.306–0.493	
Hs (1–8)	CV	1	0	1	0.000	0.000–0.793	Recent
	RV	26	10	16	0.385	0.225–0.575	
	LV	18	10	8	0.556	0.338–0.755	
	Total	45	20	25	0.444	0.309–0.588	
Ts (1–11)	CV	3	1	2	0.333	0.061–0.792	Recent
	RV	128	60	68	0.469	0.385–0.555	
	LV	110	46	64	0.418	0.330–0.511	
	Total	241	107	134	0.444	0.383–0.507	
Hk	CV	4	1	3	0.250	0.046–0.699	Recent

58

Table 12.   (cont'd—4)

Sample	Valve	N	N_R	N_Q	N_R/N	95% range	Age
Su (*1–29*)	RV	773	320	453	0.414	0.380–0.449	Recent
	LV	877	340	537	0.388	0.356–0.421	
	Total	1,650	660	990	0.400	0.377–0.424	
Is	CV	312	143	169	0.458	0.404–0.513	Recent
Km (*1–11*)	CV	4	1	3	0.250	0.046–0.699	Recent
	RV	61	20	41	0.328	0.223–0.453	
	LV	66	23	43	0.348	0.245–0.469	
	Total	131	44	87	0.336	0.261–0.420	
Sr	CV	4	0	4	0.000	0.000–0.490	Recent
	RV	1	0	1	0.000	0.000–0.793	
	LV	9	2	7	0.222	0.063–0.547	
	Total	14	2	12	0.143	0.040–0.399	
Mb	CV	1	0	1	0.000	0.000–0.793	Recent
Bt	RV	44	13	31	0.295	0.181–0.442	Recent
	LV	29	9	20	0.310	0.173–0.492	
	Total	73	22	51	0.301	0.208–0.414	
Ks	CV	4	2	2	0.500	0.150–0.850	Recent
	RV	3	1	2	0.333	0.061–0.792	
	Total	7	3	4	0.429	0.158–0.750	
Az	CV	2	0	2	0.000	0.000–0.658	Recent
	RV	1	0	1	0.000	0.000–0.793	
	LV	1	0	1	0.000	0.000–0.793	
	Total	4	0	4	0.000	0.000–0.490	
Yk	RV	2	1	1	0.500	0.095–0.905	Recent
Ec (*1–2*)	RV	417	124	293	0.297	0.255–0.343	Recent
	LV	407	122	285	0.300	0.258–0.346	
	Total	824	246	578	0.299	0.269–0.331	
Sk	CV	7	1	6	0.143	0.026–0.513	Recent
	RV	43	1	42	0.024	0.004–0.121	
	LV	39	6	33	0.154	0.073–0.297	
	Total	89	8	81	0.090	0.046–0.167	
My	RV	5	0	5	0.000	0.000–0.434	Recent
	LV	12	3	9	0.250	0.089–0.532	
	Total	17	3	14	0.176	0.062–0.410	
Kj (*1–9*)	RV	173	42	131	0.242	0.184–0.311	Recent
	LV	159	45	114	0.283	0.219–0.357	
	Total	332	87	245	0.262	0.218–0.312	
Ng	CV	10	6	4	0.600	0.313–0.832	Recent
Ta (*1–8*)	CV	1	1	0	1.000	0.207–1.000	Recent
	RV	688	176	512	0.256	0.225–0.290	
	LV	702	173	529	0.246	0.216–0.279	
	Total	1,391	350	1,041	0.252	0.230–0.275	
Hm	RV	50	13	37	0.260	0.159–0.396	Recent
	LV	65	12	53	0.185	0.109–0.296	
	Total	115	25	90	0.217	0.151–0.301	
Im	RV	38	8	30	0.211	0.111–0.364	Recent
	LV	25	7	18	0.280	0.143–0.476	
	Total	63	15	48	0.238	0.150–0.356	
Nt	CV	4	1	3	0.250	0.046–0.699	Recent
	RV	61	20	41	0.328	0.223–0.453	

Table 12. (cont'd—5)

Sample	Valve	N	N_R	N_Q	N_R/N	95% range	Age
	LV	66	23	43	0.348	0.245–0.469	
	Total	131	44	87	0.336	0.261–0.420	

are known, as discussed later. Perfect distribution overlap of this sort is quite unlikely in the case of sibling species.

3) The two phenotypes are clearly distinguishable by their different sculpture, but no significant difference can be observed as to their coloration, size, growth rate presumed from the intervals of growth rings, average number of radial ribs, various bivariate characters of shell, and anatomical characters of soft parts. This dimorphic phenomenon is therefore quite independent of the variation of other characters.

4) As a result of my examination of living populations in Sagami Bay, it was found that the breeding season is not different for the two phenotypes. The reproductive glands are fully developed in all the large specimens of both phenotypes, which were collected near the end of July (subsamples *Jg (16–21)*) and the beginning of August (subsamples *Jg (24–26)*).

Though nobody has successfully observed any living population of *C. vesiculosus* in its habitat, the ecological habits of the two phenotypes seem to be the same, judging from my own observation on freshly dredged specimens and their behavior in aquaria. As already described (p. 24), this pectinid is certainly hermaphroditic, and this discontinuous variation is never attributable to sexual dimorphism.

Such discontinuous variation of sculpture is unknown in other species of *Cryptopecten*. All the available samples of *C. bullatus*, *C. nux*, *C. phrygium*, *C. spinosus* and *C. yanagawaensis*, as well as those of *C. vesiculosus* predating the Middle Pleistocene, are monomorphic in this character, always exhibiting erect (not imbricated) scales on the interspaces of radial ribs like the Phenotype Q of *C. vesiculosus*. In this respect Phenotype R seems to represent a more specialized and probably more advanced morphology than Phenotype Q, and I regard here the former as mutant and the latter as wild-type.

Somewhat similar dimorphic phenomena can be observed in some other pectinids. As will be shown later (p. 92, Text-fig. 25), the populations of *Aequipecten commutatus* (Monterosato, 1875) from the Mediterranean seem to consist of two discrete phenotypes, the radial ribs of which may or may not have a hollow structure. In *Aequipecten opercularis* the surface of radial ribs and interspaces is either rough or smooth as with the two sub-phenotypes in Q-type individuals of *C. vesiculosus*.

The relation between *Volachlamys hirasei* (Bavay, 1904) and *Volachlamys awajiensis* (Pilsbry, 1905) is another striking case which I have observed. The surface sculpture of the two nominal species appears to be quite different; the radial ribs are rounded and much weakened in *V. hirasei*, whereas they are subquadrate and highly elevated in *V. awajiensis*. For a long time many Japanese malacologists have regarded the two phena as specifically or subspecifically distinct, but the latter was sometimes regarded as a variety or a forma of the former (Kuroda, 1932; Habe, 1951). Though individual variation has been little studied at the population level, the two phena are distributed in the same sea areas (at least in the Inland Sea between Honshu and Shikoku, and Ariake Bay of Kyushu), and are hardly distinguishable by such characters as shell size, coloration, general outline and number of radials. As a result of my preliminary survey on the collection of the National Science Museum [Tokyo], it was found that the sample NSMT 46502 labelled *V. hirasei awajiensis* and the sample NSMT 46503 labelled *V. hirasei hirasei* came from one and the same locality (Okinohata Harbor of Yanagawa City, Kyushu). Although there are a few intermediate individuals, they are basically discontinuous. Nevertheless the two samples probably came from one and the same local population. According to Dr. S. Okamura's personal communication (May 15, 1982), the two phena are also distinct but strictly sympatric at several stations in Osaka Bay. At present I am strongly inclined to regard the relation of the two phena as an example of intrapopulational discontinuous variation which is somewhat analogous to the case of *C. vesiculosus*. If this is true, *V. awajiensis* should be taxonomically treated as a junior synonym of *V. hirasei*, merely representing a sharply ribbed phenotype within one and the same species.

C. Evaluation of Ecophenotypic Effect

Of various types of ecological variation enumerated by Mayr (1969, pp. 152–155), habitat variation (or ecophenotypic effect) must be critically examined in studies of morphological variability. This effect may be defined as a purely nongenetic modification (or adjustment) of morphology in response to specific ecological conditions.

Among the various bivalves, ecophonotypic variation including xenomorphism (Stenzel, 1971, pp. N1021–1023) is predominantly known in the Ostreidae and some other groups whose shells tightly attach to hard substrates. Such ecological variation, though with less conspicuous effect, may also exist in byssate and free-living species. According to Beu (1966), for example, there are two ecophenotypes in *Chlamys dieffenbachi* (Reeve) from the southwestern Pacific; one is encrusted by sponge and the other is free-living. In this pectinid the number of radial ribs increases with growth, the increase occurring ontogenetically earlier and the shell growing much larger in the sponge-encrusted individuals than in the free-living ones. Consequently, it may be said that the number of radials varies ecophenotypically. Johnson (1981) interpreted the wide intrapopulational variation and ontogenetical flexibility of this character in the Jurassic species, *Radulopecten vagans* (Sowerby), as due to some ecophenotypic effect on the basis of analogy with the case of *C. dieffenbachi*, predicting that ecophenotypically varying organisms may be generally characterized by an ontogenetic reduction in variation.

So far as the number and mode of radial ribs of *Cryptopecten* are concerned, however, an ecophenotypic effect of this sort is unlikely in view of the following facts:

1) The radial ribs in every species of *Cryptopecten* are quite persistent except for the outermost ones, and their number never increases with growth. The variation is not flexible and seems to be inherently determined.

2) In each sample the frequency distribution of rib number is unimodal, resembling a normal curve.

3) Judging from the dredged living shells of *C. vesiculosus* in Sagami Bay, there is no evidence for ecological seqregation of the two phenotypes; their mode of life as well as attached organisms seem to be quite the same.

4) The phenotypic frequency is almost stable in the same sea area.

For the same reasons, I believe that other non-genetic causes (e. g., climatic, host-determined, density-dependent and traumatic) are improbable for the sculpture dimorphism of *C. vesiculosus*.

CHAPTER 8

Geographic Variation of *Cryptopecten vesiculosus*

Geographic variation is, so to speak, spatial integration of intrapopulational variation, and evolutionary change within a lineage is chronological integration of geographic variation. Quantitative examination of geographic variation seems to be indispensable for the study of phyletic evolution as well as speciation. Gould and Johnson (1972) emphasized the importance of the study of geographic variation and the potential role of multivariate analyses which can treat the variation of many characters (morphological, physiological, biochemical, etc.) simultaneously. In the present article, however, I hesitate to apply this method to the morphometric data of *Cryptopecten*, chiefly because causal evaluation of various unit characters is insufficient. Some characters may represent genetic difference between populations, while other characters may be strongly influenced by environmental factors. Various kinds of heterogeneity among samples also precludes this approach. Consequently, I intend to clarify the morphological differences among population samples on the basis of each univariate or bivariate character and then to approach synthetically the general aspect of geographic variation.

Ideally, geographic variation should be determined from many large population samples covering the whole distribution of the species in question. Among the three living species of *Cryptopecten*, reliable conclusions may be obtained only for *C. vesiculosus*, whose distribution is virtually restricted to the seas around the central and southwestern part of the Japanese Islands. Some qualitative observations were made also on the geographic variations of *C. bullatus* and *C. nux*, which distribute more extensively than *C. vesiculosus*, and the results will be shown in the section of systematic description.

In spite of recent progress in the chronology of Neogene and Quaternary marine deposits in Japan, the geographic variation of *C. vesiculosus* in the geological ages is still difficult to clarify. More than 15 large Recent samples of this pectinid, including more than 100 subsamples, are available for the study of geographic variation, though the material is still not necessarily ideal. The localities of large samples concentrate on the Pacific coast of central Honshu (Kanto and Tokai Regions) and the western coast of Kyushu, and are rather sparsely situated in the Japan Sea and the seas surrounding the Ryukyu Islands. The well-preserved large Holocene fossil sample *Ms* (only about 2,720 years B. P. according to Matsushima's ^{14}C dating) from south Kyushu may also be useful for the study of geographic variation.

1. Shell Size and Growth Rate   [Table 13]

Generally speaking, body size is regarded as an easily measurable character that varies geographically. In the present case, however, there are some problems, because the age (and size) composition of the samples may be quite variable. The height of the maximum individual in each sample (H_{max}) may be informative, but it is certainly in-

62

Table 13. Maximum shell size and average growth amount in some selected samples of *Cryptopecten vesiculosus*.

Sample	N	N_g	H_{max} (mm)	$\bar{H}_1$ (mm)	s (mm)	V	O. R. (mm)
Sh 1	229	0	34.5	—	—	—	—
Nj	29	0	36.3	—	—	—	—
Ik	154	35	34.4	21.320 ± 0.818	4.840	22.70	13.3–29.5
Kg 1	20	7	43.4	27.200 ± 2.745	7.263	26.70	14.1–35.4
Ob	85	10	35.9	18.120 ± 1.843	5.828	32.16	10.7–26.8
Tm 1+2	36	8	35.3	18.250 ± 1.796	5.079	27.83	10.8–23.6
Sn 3	177	117	35.0	19.454 ± 0.447	4.836	24.86	8.4–28.1
Ab 1	713	420	37.4	19.089 ± 0.219	4.479	23.46	8.0–27.2
Sm	250	16	40.7	14.363 ± 0.592	2.369	16.50	12.0–20.6
Ms	1,410	468	31.7	14.711 ± 0.136	2.933	19.94	8.7–24.4
Hy	352	22	25.9	15.841 ± 0.812	3.808	24.04	10.8–22.3
Jg (8)	1,838	248	29.5	15.340 ± 0.245	3.852	25.11	8.0–24.1
Ts (11)	116	62	33.2	19.326 ± 0.442	3.481	18.01	9.8–29.0
Su (29)	493	219	34.0	15.079 ± 0.338	3.778	25.06	8.7–24.7
Is	312	140	29.7	15.104 ± 0.203	2.302	15.24	10.2–22.2
Km (1–11)	131	26	30.7	14.915 ± 0.768	3.914	15.32	8.2–21.0
Ec 2	436	65	28.9	15.154 ± 0.497	4.009	26.46	8.7–23.8
Ta 2	318	96	27.5	14.028 ± 0.420	4.119	29.36	8.2–22.6
Hm	115	52	30.3	16.679 ± 0.351	2.534	15.19	10.6–23.6

N: total number of individuals; N_g: number of individuals possessing growth ring(s); H_{max}: overall height of shell in the maximum individual in each sample; $\bar{H}_1$: average height of shell at the first growth ring; s: standard deviation of H_1: V: coefficient of variation of H_1: O. R.: observed range of H_1.

fluenced by sample size. In this respect the average size (height) at the first stepwise growth ring ($\bar{H}_1$) may be more meaningful, although the intrapopulational variation of this character is generally wide.

On the basis of analysis of several large Recent samples of *C. vesiculosus*, it can be said that the geographic variation of size, represented by the difference of the values of H_{max} and $\bar{H}_1$ is rather inconspicuous. More precisely, Recent individuals rarely exceed 30 mm in overall height, and in most cases the values of $\bar{H}_1$ are confined to the range from 14 mm to 17 mm. In the one exceptional case of sample *Ts (11)*, $\bar{H}_1$ exceeds 19 mm. Although the sample size is generally small, many subsamples of *Zs*, *Hs* and *Ts* from the three sea banks off the Izu-shichitô Islands also show large maximum size and rapid growth rate in comparison with Recent samples from other sea areas. I presume that conditions for shell growth are optimum on these sea banks, where oceanic water prevails. Some Recent large samples (e. g., *Kj (1–9)*) are almost entirely composed of small individuals without any growth ring, but they do not seem to indicate any significant geographic variation.

2. Shape of Shell (Bivariate morphometric characters) [see Tables 4–10]

The results of various bivariate analyses on seven Recent samples and one Holocene fossil sample seem to offer data for the consideration of geographic variation. The allometric relations between L and H, L and T, and L and D as well as standardized form ratio H/L, T/L and D/L in these samples were compared. Although these bivariate characters

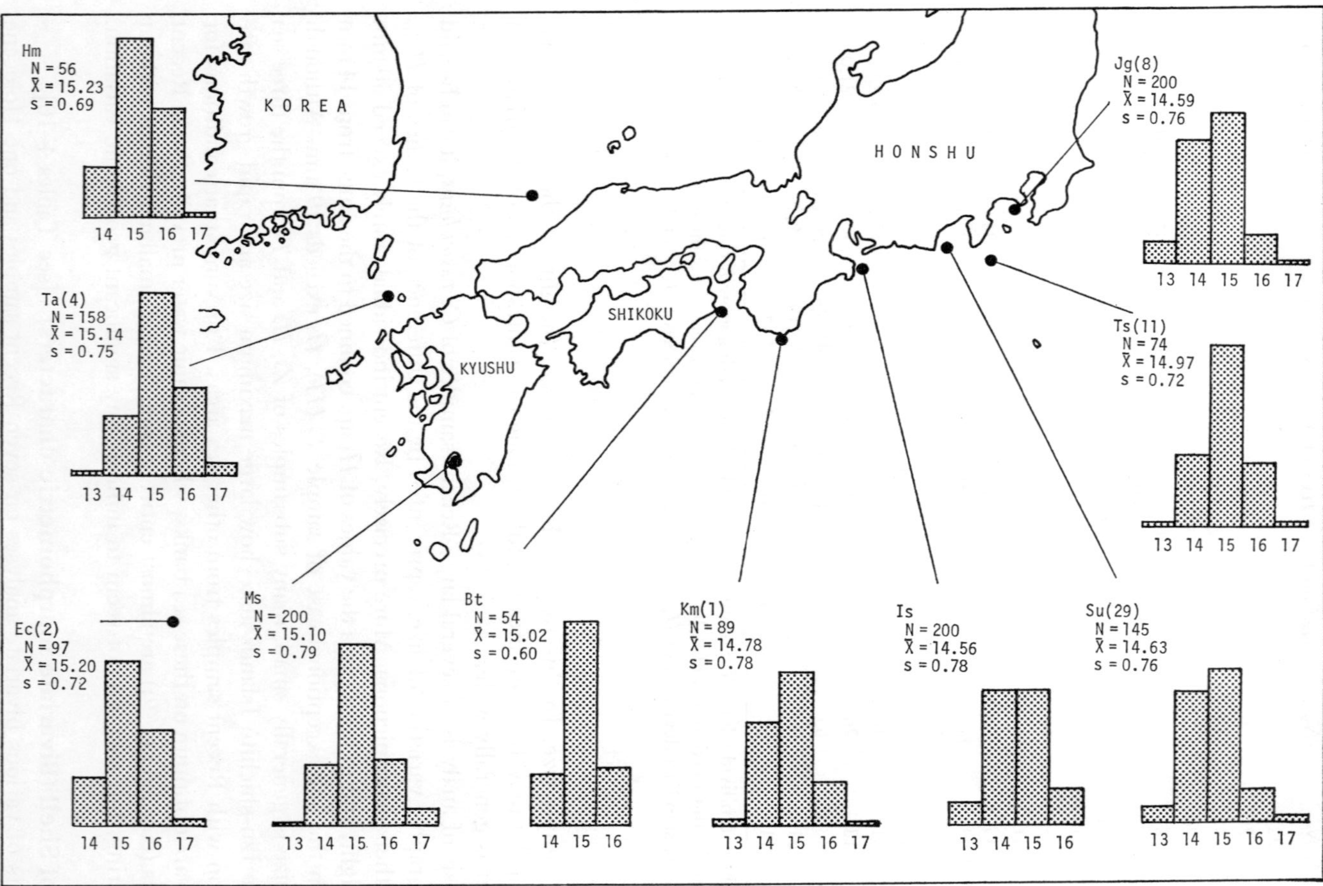

Fig. 20. Geographic variation of the average number of radial ribs in selected Recent samples of *Cryptopecten vesiculosus*. *N*: number of examined individuals, *X̄*: average number of radial ribs, *s*: standard deviation. *Jg* (*8*): off Jogashima Islet, *Ts* (*11*): Takase, *Su* (*29*): Senoumi, *Is*: off Anori, *Km* (*1*): off Kushimoto, *Bt*: off Benten Islet, *Ms*: Moeshima Islet (subfossil), *Ec* (*2*): west of Tokara Islets, *Ta* (*4*): south of Tsushima Islands, *Hm*: off Hamada. Presence of a cline is strongly suggested in the populations along the Pacific coast of southwest Japan.

Table 14. Results of *F* test and *t* test indicating the significance for the differences of variance and mean of number of radial ribs between Recent samples.

Sample		Sample										
		Pacific Coast of Honshu							E. China Sea		Japan Sea	
		Hy	*Jg (8)*	*Ts (11)*	*Su (29)*	*Is*	*Km (1)*	*Bt*	*Ms*	*Ec (1)*	*Ta (4)*	*Hm*
	Hy	*	N	N	N	N	N	N	N	N	N	N
	Jg (8)	1.37	*	N	N	N	N	S	N	N	N	N
Pacific	*Ts (11)*	2.09	3.81	*	N	N	N	N	N	N	N	N
Coast	*Su (29)*	0.83	0.59	3.17	*	N	N	S	N	N	N	N
of	*Is*	1.58	0.33	3.97	0.88	*	N	S	N	N	N	N
Honshu	*Km (1)*	0.43	1.95	1.67	1.36	2.16	*	S	N	N	N	N
	Bt	2.42	3.89	0.37	3.35	4.01	1.97	*	S	N	S	N
E. China	*Ms*	3.55	6.60	1.16	5.44	6.82	3.20	0.66	*	N	N	N
Sea	*Ec (2)*	4.24	6.63	2.01	5.76	6.76	3.84	1.54	1.07	*	N	N
Japan	*Ta (4)*	3.97	6.88	1.59	5.80	7.08	3.60	1.06	0.54	0.60	*	N
Sea	*Hm*	4.01	5.75	2.07	5.12	5.84	3.59	1.73	1.18	0.30	0.81	*

Right upper: Result of *F* test [N: difference of variance not significant with 95% confidence; S: difference of variance significant with 95% confidence but not with 99% confidence].

Left lower: Result of *t* test [Significant $(0.05 > P > 0.01)$ and very significant $(0.01 > P)$ differences indicated by single and double underlining, respectively].

sometimes differ significantly between samples, no conspicuous tendency of geographic variation is detected here.

3. Number of Radial Ribs   [see Tables 11 and 14]

As mentioned before, the frequency distribution of the number of radial ribs on the disk (X) is unimodal and essentially normal in every sample, the mode lying at 14 or 15 in all Recent samples. However, the mean $(\bar{X})$ is often significantly different between samples of different sea areas. It is as small as 14.5–14.8 in the samples from the Pacific coast of central Honshu, while it is commonly larger than 15.0 in the samples from the East China Sea and Japan Sea (Fig. 20). The Holocene fossil sample *Ms* from south Kyushu seems to belong to the latter cluster in this character. The sample *Ts (11)* and some other small samples from three sea banks off the Izu-shichitô Islands, and the sample *Bt* from Kii Strait between Kii Peninsula and Shikoku are actually intermediate between the two clusters. The results of *F* test and *t* test are collectively shown in Table 14. These data seem to indicate that the geographic variation of this character constitutes a gentle cline along the Pacific coast of the Japanese Islands; the average number of radial ribs becomes more numerous towards the warm southwest seas. This clinical change, however, is not apparent along the coasts of the East China Sea and the Japan Sea.

4. Coloration of Shell

The geographic variation of color pattern and ground color is intuitively not much different among Recent samples. Every large sample contains yellow-orange monotonic

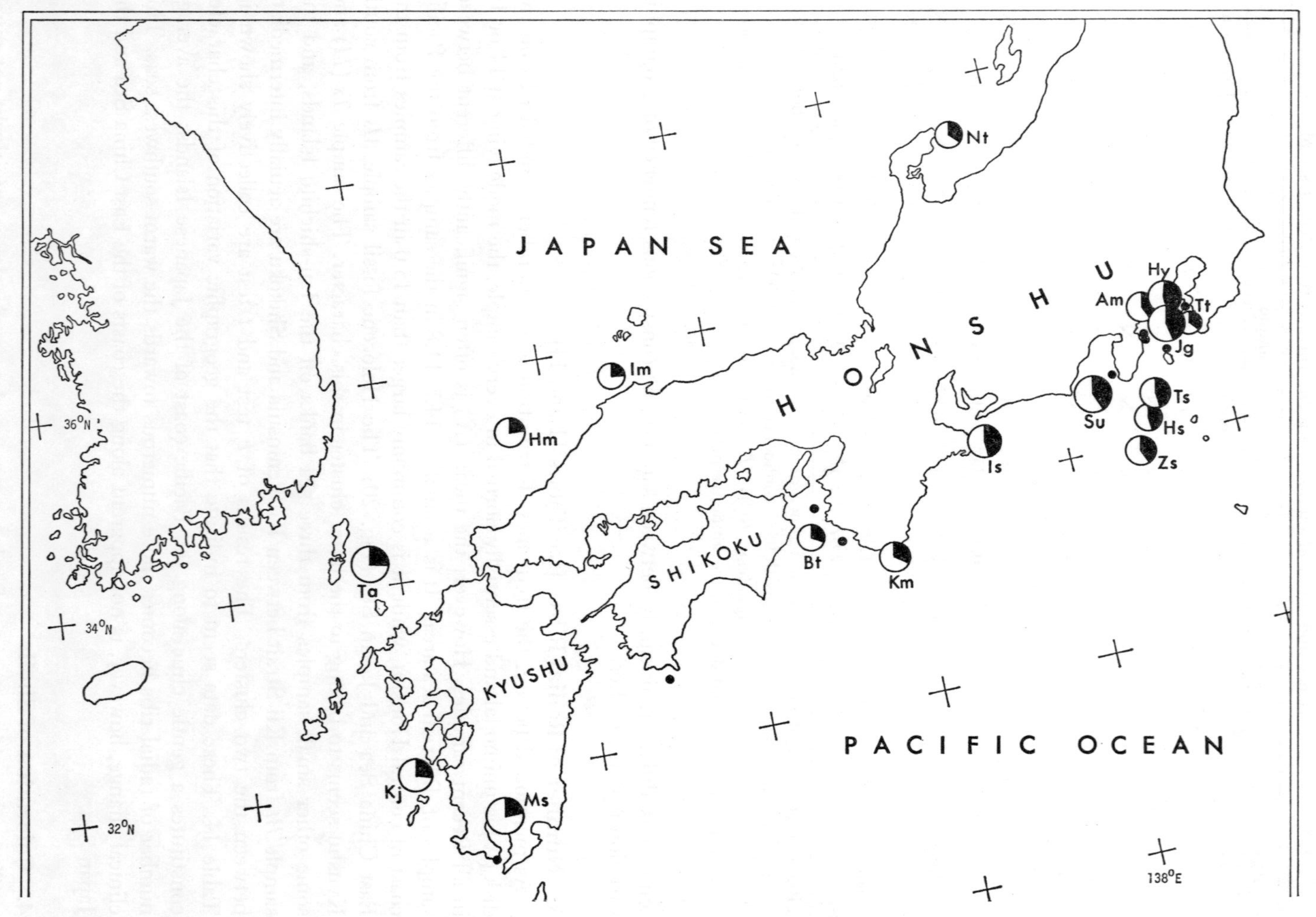
JAPAN SEA
Nt
HONSHU
Hy
Am
Tt
Jg
Im
Su
Ts
Hs
Hm
Zs
Is
Bt
SHIKOKU
Km
Ta
36°N
34°N
32°N
KYUSHU
Kj
Ms
PACIFIC OCEAN
138°E

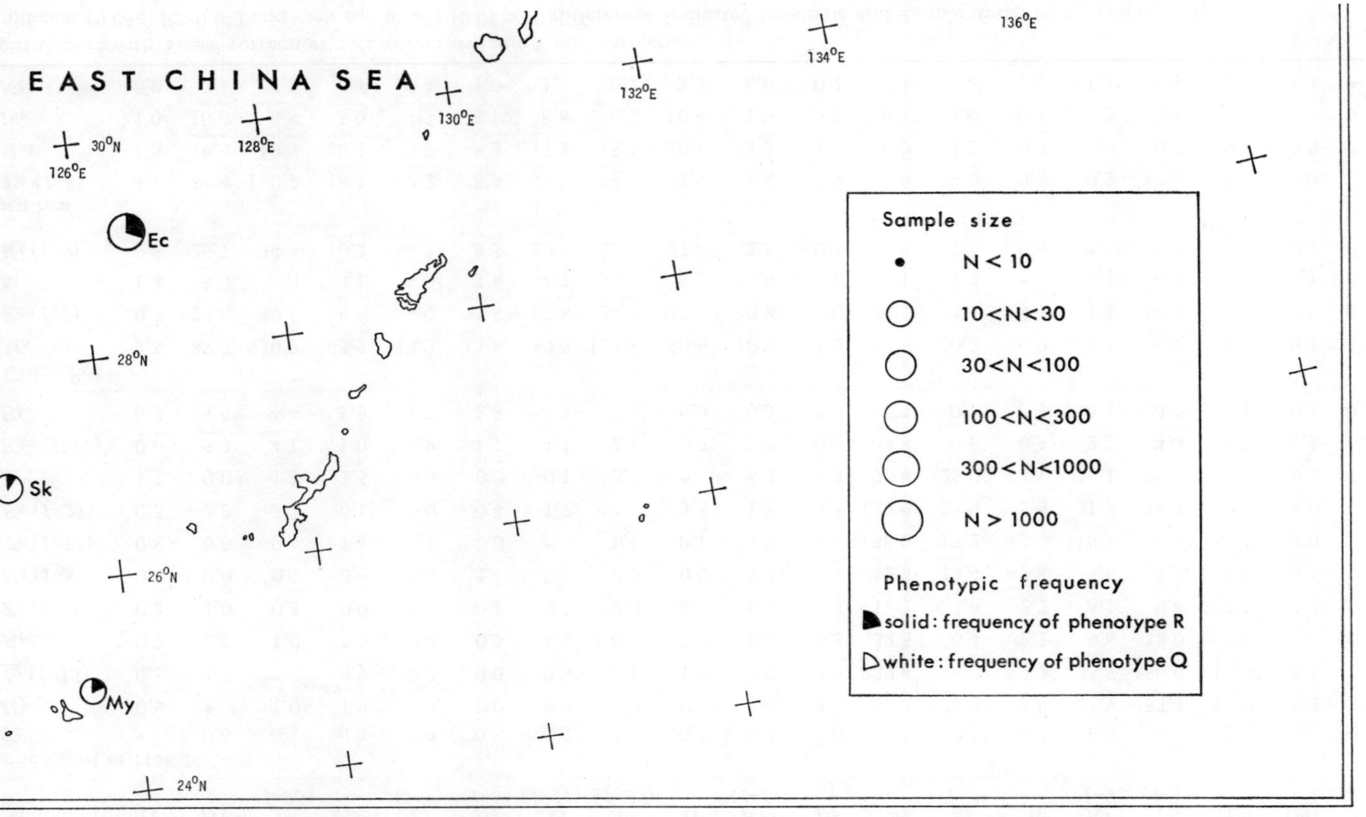

Fig. 21. Geographic variation of the phenotypic frequency in selected Recent samples of *Cryptopecten vesiculosus*. All the subsamples are integrated. *Tt*: Tateyama Bay, *Hy*: off Hayama, *Am*: Amadaiba, *Jg*: off Jogashima Islet, *Ts*: Takase, *Hs*: Hyotanse, *Zs*: Zenisu, *Su*: Senoumi, *Is*: off Anori, *Km*: off Kushimoto, *Bt*: off Benten Islet, *Ms*: Moeshima Islet, *Ec*: off Tokara Islets, *My*: off Miyako Island, *Sk*: off Senkaku Islets, *Kj*: East China Sea west of Kyushu, *Ta*: Tsushima Strait, *Hm*: off Hamada, *Im*: off Izumo Peninsula, *Nt*: off Suzu. Presence of a cline is strongly suggested in the populations along the Pacific coast of southwest Japan.

Table 15. Chi-square matrix indicating the significance for the difference of phenotypic frequency between Recent samples (2×2 contingency tables).

| Sample | | Sample | | | | | | | | | | | | | | | | | |
| | | Pacific Coast of Honshu | | | | | | | | | | East China Sea | | | | Japan Sea | | | |
Sample	Tt	Hy	Jg (1–15)	Am	Zs (1–9)	Hs (1–8)	Ts (1–11)	Su (1–29)	Is	Km (1–11)	Bt	Ms	Ec (1–2)	My	Kj (1–9)	Ta (1–8)	Hm	Im	Nt
Pacific Coast of Honshu																			
Tt	*	0.6	0.3	0.0	0.0	0.3	0.5	0.1	0.7	0.0	0.0	1.7	0.1	0.7	0.4	0.7	1.2	0.6	0.0
Hy	0.9	*	1.0	1.9	0.8	0.0	0.0	3.0	0.0	4.8	5.0	82.2	25.0	3.9	25.9	53.4	19.0	9.2	1.7
Jg (1–15)	0.5	1.1	*	0.9	0.2	0.0	0.3	3.1	1.4	3.7	3.9	229.4	48.5	2.2	33.7	150.0	19.0	8.1	1.0
Am	0.2	2.1	1.0	*	0.0	0.3	1.2	0.0	2.2	0.8	1.5	32.5	6.4	2.2	9.6	18.0	9.5	4.3	0.2
Zs (1–9)	0.2	1.0	0.3	0.0	*	0.1	0.5	0.0	0.7	0.7	1.3	17.1	3.6	2.2	6.1	9.4	7.3	3.7	0.2
Hs (1–8)	0.6	0.0	0.1	0.5	0.3	*	0.0	0.2	0.0	1.3	1.9	12.3	13.4	2.8	5.6	7.5	7.2	4.2	0.7
Ts (1–11)	0.8	0.0	0.4	1.4	0.7	0.0	*	1.5	0.1	3.7	4.1	58.0	17.2	3.6	19.8	36.8	16.2	8.0	1.3
Su (1–29)	0.3	3.2	3.2	0.1	0.0	0.4	1.7	*	3.5	1.8	2.5	122.9	23.9	2.6	21.8	74.3	14.3	6.0	0.5
Is	1.1	0.0	1.5	2.5	0.9	0.0	0.1	3.7	*	5.2	5.3	79.4	25.0	4.1	26.1	51.9	19.4	9.5	1.9
Km (1–11)	0.0	5.3	4.1	1.0	0.9	1.7	4.1	2.1	5.7	*	0.1	9.8	0.6	0.5	2.2	4.0	3.7	1.5	0.0
Bt	0.2	5.6	4.4	1.9	1.7	2.5	4.7	2.8	6.0	0.3	*	2.7	0.0	0.5	0.3	0.7	1.3	0.4	0.5
E. China Sea																			
Ms	2.5	83.5	230.2	33.5	18.2	13.6	59.2	123.8	80.6	10.5	3.2	*	20.3	0.0	3.5	5.7	0.0	0.1	3.6
Ec (1–2)	0.3	25.6	49.1	6.8	4.0	14.8	17.8	24.4	25.7	0.8	0.0	20.7	*	0.7	1.4	13.6	2.9	0.8	0.2
My	1.4	5.0	3.0	3.1	3.0	3.8	4.7	3.5	5.2	0.9	1.1	0.1	1.2	*	0.3	0.2	0.0	0.1	0.9
Kj (1–9)	0.8	26.7	34.4	10.2	6.7	6.5	20.6	22.4	27.0	2.5	0.5	3.8	1.5	0.6	*	0.1	0.7	0.1	0.9
Japan Sea																			
Ta (1–8)	1.1	54.4	150.7	18.7	10.2	8.5	37.7	74.9	52.9	4.4	0.9	5.9	14.0	0.5	0.2	*	0.5	0.0	1.5
Hm	1.8	19.9	19.8	10.2	8.2	8.3	17.1	15.1	20.4	4.3	1.7	0.0	3.2	0.2	0.9	0.7	*	0.0	2.1
Im	1.0	10.0	8.8	5.0	4.3	5.1	8.8	6.7	10.4	1.9	0.7	0.2	1.0	0.3	0.2	0.1	0.1	*	0.9
Nt	0.0	2.1	1.3	0.4	0.4	1.0	1.7	0.7	2.3	0.0	0.2	4.4	0.4	1.6	1.3	1.9	2.7	1.4	*

Right upper: with Yates' correction; left lower: without Yates' correction.
Significant [0.05>P>0.01] and very significant [0.01>P] differences indicated by single and double underlining, respectively.

individuals and shows distinct color polymorphism. The phenotypic frequency is, however, commonly less than 5 percent and too unreliable to be applied as an indicator of geographic variation.

5. Surface Sculpture (**Phenotypic frequency**)   [see Tables 12 and 15]

The two discrete phenotypes, Phenotypes Q and R, are clearly separable in every Recent sample, and no individual is intermediate. The relative frequency of Phenotype R as well as a range of 95 percent confidence (obtained by small sample method) was calculated in each sample on the basis of all the constituent individuals.

In the previous study (Hayami, 1973), I concluded by means of chi-square tests that no significant difference of phenotypic frequency exists between any pair of Recent samples, and that the living populations of *C. vesiculosus* may be considerably panmictic. The samples treated at that time, however, are from only a limited sea area, the Pacific coast of central Honshu. Also I was uneasy about the variation of the Holocene fossil sample *Ms*, whose phenotypic frequency is unexpectedly low for its very young geological age.

As a result of the present study on samples from much wider sea areas, it was found that phenotypic frequency is often significantly different between samples of distant areas, particularly between samples from the Pacific coast of central Honshu and samples from the East China Sea–Japan Sea side of the Japanese Islands, as documented by a newly prepared chi-square matrix (Table 15).

All the large samples from the Pacific coast of central Honshu (e. g., *Hy*, *Jg*, *Am*, *Zs*, *Ts*, *Su* and *Is*) show nearly the same phenotypic frequency within a range from 0.39 to 0.46, whereas in the samples from the East China Sea and Japan Sea (e. g., *Sk*, *My*, *Ec*, *Kj*, *Ta*, *Hm* and *Im*) the frequency is much lower, ranging from 0.09 to 0.30. Therefore, the Holocene fossil sample *Ms* from south Kyushu is not in the least unusual with respect to this character. The samples *Km* and *Bt* off Kii Peninsula appear to be intermediate between the two clusters as is their geographic location (Fig. 21).

The geographic variation of this character also indicates the presence of an evident cline along the Pacific coast of the Japanese Islands. The available data for the East China Sea and Japan Sea do not show any conspicuous geographic variation. However, the northernmost sample *Nt* off Noto Peninsula, though of relatively small size, shows considerably high phenotypic frequency, and therefore it is probable that somewhat similar clinal change also exists along the Japan Sea side. It is remarkable that the conclusion on the geographic variation of this character roughly coincides with that on the average number of radial ribs.

6. Causal Evaluation of Geographic Variation

The genetic background for the variation of these characters is still unknown. As it dwells only in environments under the influence of oceanic water, Mendelian experiments on this pectinid are probably more difficult than on molluscs in coastal waters. Circumstantial evidence about the intrapopulational and geographic variations, however, convinces me that some of these unit characters are strongly controlled by genetic factors. Among others, the number of radial ribs on the disk and the relative frequency of phenotype *R*, both of which show obviously clinal changes at least along the Pacific side of the southwestern part of the Japanese Islands, are most striking. The number of ribs is probably polygenic in view of its normal-type frequency distribution and uniformity of variability in each sample. The dimorphism is, on the other hand, probably determined by a single (or a few) genetic factor such as an allele or chromosomal aberration, as I presumed somewhat boldly before (Hayami, 1973).

As discussed by Mayr (1963, p. 361), clines are generally the products of two conflicting forces: one is the differential selection pressure in accordance with local environments (or environmental gradient), and the other is gene flow among local populations. The former acts to increase the difference of gene frequency among local populations, while the latter contributes to homogenizing the gene frequency throughout the distribution of the species in question. Because the present pectinid is an open sea dweller, a considerable amount of gene flow can be expected during planktic larval stages. Water temperature evidently decreases from southwest to northeast on both sides of the Japanese Islands and may be one of the significant factors causing the environmental gradient. At least on the Pacific side, the clinal change of phenotypic frequency as well as average number of radial ribs seems to correlate well with the gradient of water temperature. It is, however, difficult to say what environmental factors are actually responsible for the differential selection pressure, because, as pointed out by several scientists, significant correlation does not necessarily mean direct causal relation.

CHAPTER 9

Phyletic Evolution of *Cryptopecten vesiculosus*

The fossil records of *Cryptopecten vesiculosus* can be traced back to the Middle Pliocene (ca. 3.5 Ma). Though some dead-end incipient species may have arisen from this stock, many fossil samples seem to indicate the persistency and phyletic morphological change of this species. The geographic distribution of this pectinid, especially the northern and southern limits, must have fluctuated during the geologic past in accordance with changing marine conditions, but its center seems to have remained in the seas surrounding the Japanese Islands. So far as I am aware, there is no undoubted fossil occurrence of *C. vesiculosus* in countries other than Japan.

More than 10 Pliocene, more than 20 Pleistocene and one Holocene fossil samples in addition to a number of Recent ones are available for the study of phyletic evolution of this species (see List of Examined Samples (pp. 127–137) for their localities, horizons, etc.), though some of them are not very useful owing to the small sample size and ambiguous geologic age. Many characters of these fossil samples, except for some Late Pleistocene and Holocene ones, are clearly outside the range of geographic variation in the present seas, and are mainly attributable to evolutionary change. At the same time it is here recognized that some characters changed directionally with time.

1. Shell Size and Growth Rate   [see Table 13]

The overall shell height of the largest individual (H_{max}) and the average shell height at the formation of the first stepwise growth ring ($\bar{H}_1$) for several selected large fossil and Recent samples were compared. The value of H_{max} is inevitably influenced by sample size, but in this case a large sample does not necessarily show a large value of H_{max}. It may be concluded that the ultimate size of gerontic individuals generally decreases with time. Large individuals over 30 mm are commonly found in every Pliocene and Pleistocene sample, while even gerontic individuals with many growth rings seldom exceed 30 mm in Holocene and Recent samples.

The phyletic size decrease is also clearly indicated by the different values of $\bar{H}_1$. The variation of $\bar{H}_1$ is actually so wide that the coefficient of variation (V) ranges from 15 to 30 in these selected samples. In the two Pliocene samples *Sh 1* and *Nj* (and probably also *Sh 2*), growth rings are, if present, rather weak, notwithstanding the fact that the shell often attains a size of over 30 mm. It is not known, however, whether this phenomenon too is attributable to phyletic change or simply due to different environmental conditions. Two other Late Pliocene samples, *Ik* and *Kg 1*, and all the Pleistocene samples contain a number of large individuals with clear stepwise growth rings. These fossil samples, except for the youngest one, *Sm*, generally have much larger values of $\bar{H}_1$ than Recent samples. Therefore, not only the ultimate size but also the annual growth rate must have decreased with time. Though the correlation between the absolute age and $\log_e \bar{H}_1$ is not

71

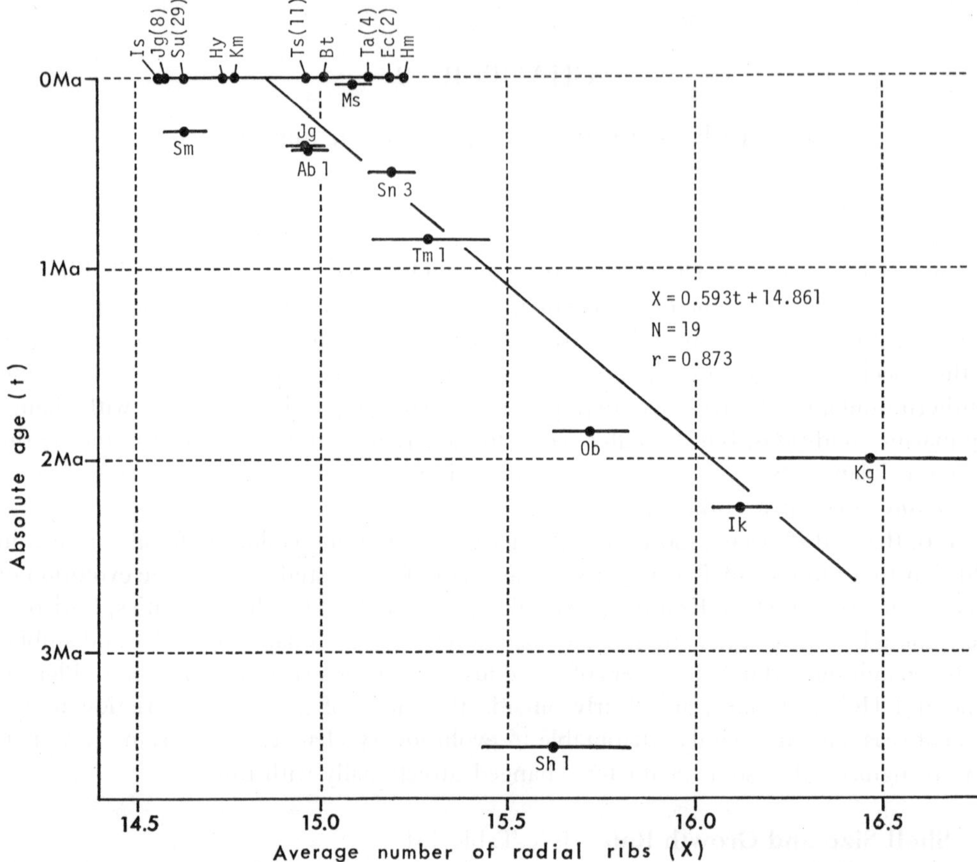

Fig. 22. Chronological change of the average number of radial ribs in the stock of *Cryptopecten vesiculosus*. Horizontal bars indicate standard errors. Regression line (*X* on *t*) on the basis 19 post-Middle Pliocene samples, indicating phyletic decrease of rib number. [Erratum: read Jz for Jg]

very high ($N=10$, $r=0.659$, $0.01 < P < 0.05$), the calculated slope of the regression line by the least square method indicates the evolutionary rate of this size decrease to be approximately 120 millidarwins.

The Late Pliocene sample *Kg 1* shows unusually larger values of H_{max} and $\bar{H}_1$ than other Pliocene and Pleistocene samples. Individuals over 40 mm are commonly found in this sample. As is discussed and described in later pages, however, some taxonomic distinction may be required for this sample in view of the unusual mode of fossil occurrence and some other unique morphological characteristics.

2. Shape of Shell [see Tables 4–10]

The allometric equations of average relative growth between *L* and *H*, *L* and *T*, and *L* and *D*, as well as standardized form ratios *H/L*, *T/L* and *D/L*, were obtained on several fossil samples. The values of growth ratio (*a*) are sometimes significantly different between samples of different ages, but there is nothing like a definite trend.

The Pliocene sample *Ik* seems to show unusually smaller values of H/L (i. e., comparatively low outline) in every growth stage than other contemporary and younger samples (Plate 7, Figs. 9, 10). In contrast, the Middle Pleistocene sample *Tm 1* (and also *Tm 2*) has decidedly larger values (i. e., comparatively tall outline) in comparison with other samples (Plate 7, Figs. 7, 8). These peculiar features are so conspicuous that one can perceive their uniqueness at a glance. However, no definite trend with time can be detected for this character and other form ratios.

3. Number of Radial Ribs [see Table 11]

The number of radial ribs on the disk (X), particularly its average and variability, is an easily recognizable and probably meaningful character on which chronological change can be discussed. This character was examined on five large Pliocene and seven large Pleistocene samples. These fossil samples show similarly unimodal and nearly normal frequency distribution in the number of radial ribs and comparable coefficient of variation (V) with Recent samples. Taking the aforementioned data on Recent samples into consideration, it is concluded that the average number of radial ribs ($\bar{X}$) has decreased considerably with time. The obtained values of $\bar{X}$ in all the examined Pliocene samples (*Sh 1, Nj, Ik, Kg 1* and *Mz 2*), which sometimes exceed 16.0, are evidently larger than those of the Pleistocene and later samples. The Early and Middle Pleistocene samples (*Ob, Tm 1, Sn 3*) seem to be intermediate in this character between Pliocene and Holocene-Recent ones. The values of $\bar{X}$ of the Late Pleistocene samples (*Ab 1, Jz, Sm*), which are all from the Pacific coast of central Honshu, are within the range of the geographic variation in the present seas, but usually significantly larger in comparison with Recent samples of the same region.

As illustrated in Fig. 22, the obtained values of $\bar{X}$ in these samples and their estimated absolute ages show a significant correlation ($N=18$, $r=0.873$, $P<0.01$). Among the Late Pliocene samples, *Kg 1* (and also *Kg 2*) has a somewhat unusually large value, and is taxonomically separable from other samples in view of some other morphological and paleoecological characteristics. If this sample is excluded, the correlation coefficient becomes still higher. This is therefore regarded as an example of a directional change of phyletic evolution which took place during geological ages.

The geographic variation of $\bar{X}$ is, as already described, considerably wide in Recent populations. Judging from the historical change of this character, the living populations on the Pacific side of central Honshu may be regarded as more advanced than those in the East China Sea and the Japan Sea. All the examined samples from the latter sea areas (*Ec 2, Ta 4, Hm*) are comparable in this character with the Middle Pleistocene fossil samples (*Tm 1, Sn 3*) of the former area.

The oldest sample of *C. vesiculosus*, *Sh 1*, shows a value of $\bar{X}$ somewhat smaller than other Pliocene samples. Though the fossil records of this stock become very obscure before the Late Pliocene, this sample may suggest that phyletic change was not necessarily unidirectional in Pliocene times.

4. Surface Sculpture (Phenotypic Frequency) [Table 12]

Discrimination between the two phenotypes Q and R is also possible in every fossil sample, even those in poorly preserved condition. Therefore, the phenotypic frequency

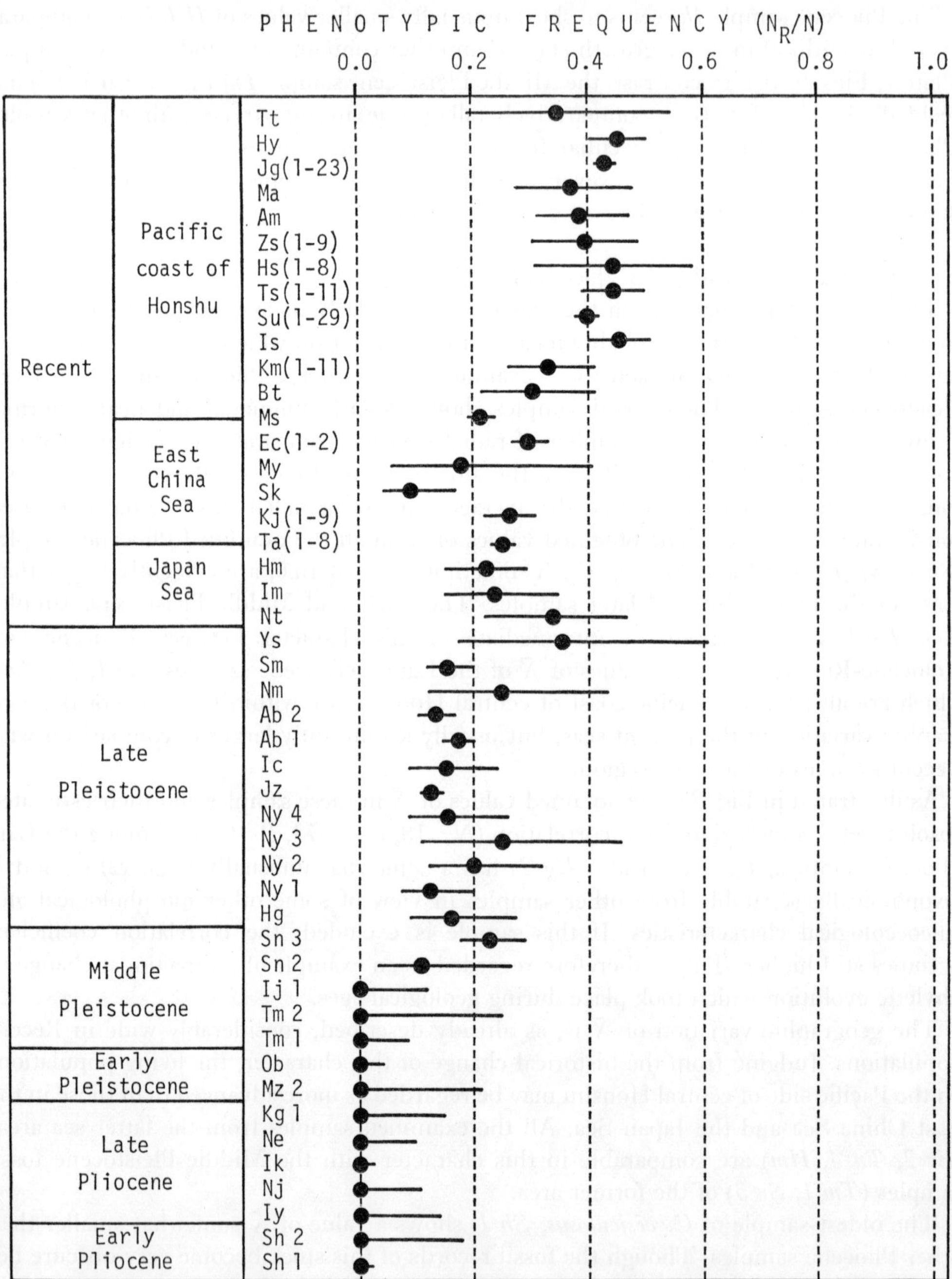

Fig. 23. Chronological change of phenotypic frequency in the stock of *Cryptopecten vesiculosus*. N: total number of individuals, N_R: number of individuals belonging to Phenotype R. Horizontal bars indicate the ranges of 95 percent confidence. See also Table 11.

N_R/N is a clean-cut and useful character for the study of phyletic change at the population level.

The historical change of this character was discussed in some detail in my previous paper (Hayami, 1973). My conclusion there about the gradual change of phenotypic frequency as well as the concept of phenotypic substitution as a cause of morphological break without any intermediate form (Hayami and Ozawa, 1975) is fundamentally supported by subsequently accumulated data.

Many fossil samples of *C. vesiculosus* were examined, but no individual belonging to the Phenotype *R* could be found in Pliocene and Early Pleistocene samples. The Middle Pleistocene samples *Ij 1* and *Ij 2* are also entirely composed of individuals belonging to the Phenotype *Q*. The earliest occurrence of the Phenotype *R* so far known on the Pacific side is from the Middle Pleistocene Sanuki Formation (samples *Sn 1, 2, 3*, all approximately 0.5 Ma). Though the observed phenotypic frequency is somewhat variable among these Sanuki samples, the largest sample, *Sn 3*, shows a value of N_R/N as high as 0.23. I have found a few individuals of the Phenotype *R* in a small sample (Dr. Ogasawara's private collection) from the Middle Pleistocene Shibikawa Formation of Oga Peninsula, signifying the nearly coeval appearance of this phenotype in populations on the Japan Sea side.

In the Late Pleistocene, a temporal transgression with strong influence of warm current occurred, leaving prolific fossil beds of "Jizôdô fauna" in the middle part of the Boso Peninsula. The samples *Ny 1, 2, 3, 4, Jz*, and *Ab 1, 2* are attributable to this horizon (approximately 0.37 Ma). These samples often show somewhat lower phenotypic frequency than the Middle Pleistocene Sanuki samples, but the value of N_R/N is also considerably variable within the Jizôdô Formation. For example, the two large samples *Ab 1* and *Ab 2*, which were randomly collected from two different layers about 2 meters apart of one and the same outcrop at Atebi, show significantly different phenotypic frequency; the sample *Ab 1*, taken from the lower layer, is richer in the Phenotype *R* than the sample *Ab 2*, taken from the upper layer. Within this brief period, therefore, minor fluctuation of phenotypic frequency may have occurred. The observed values of N_R/N in the large samples of this horizon are, however, confined to the range from 0.12 to 0.20, and it is probably appropriate to roughly estimate the average phenotypic frequency at the Jizôdô transgression is at 0.15.

The horizon of the sample *Sm* was actually determined by Sugihara et al. (1978), using marker tephras to be the upper part of the Yabu Formation (approximately 0.29 Ma). The value of N_R/N of this sample is not much different from the average value of the Jizôdô samples.

The geographic variation of the phenotypic frequency in Recent samples is, as stated before, considerably wide. However, large Recent samples from the Pacific coast of central Honshu seem to be fairly stable in this character; the observed frequency ranging from 0.39 to 0.46 is much higher in comparison with these Pleistocene fossil samples. The phenotypic frequency in the Recent samples from the East China Sea and Japan Sea is as low as from 0.17 to 0.30, but still seems to be higher than that of the Pleistocene ones. Because there is no large Middle or Late Pleistocene sample in the areas facing the East China Sea and Japan Sea, almost nothing has been known as to the geographic variation of this character in the past. As shown in the collective illustration in Fig. 23, how-

76

ever, the chronological change of the phenotypic frequency may be said to have a trend, if not truly unidirectional.

The present conclusion is unexpectedly in harmony with the aforementioned phyletic decrease in the average number of radial ribs. This may not mean pleiotrophic effect of the same genetic factor, but it is not surprising to see that the phyletic change of the two independent characters, and probably also the related gene frequency, are the most advanced in the peripheral area of distribution (Pacific coast of central Honshu). The populations in the East China Sea and Japan Sea may be more conservative, the genetic nature having perhaps changed somewhat behind the Pacific side populations.

In 1973, I estimated the values of selection coefficient (s) as well as those of the change of gene frequency per one generation (Δq) based on an assumption of the simplest relation between genotypes and phenotypes. The genetic background of this dimorphism is still unknown, but, on the same assumption, the estimated values of s and Δq should be revised, because recent geochronology reveals that the absolute age of the Jizôdô Formation was greatly underestimated in my 1973 article. Assuming the same time span of one generation (two years on average), the following revised values are obtained:

1)   In the case of dominant mutant gene [R-type individuals to be dominant homozygotes plus heterozygotes]

$s=8.6\times10^{-6}$

$\Delta q=5.7\times10^{-7}$ for fossil populations of Jizôdô Formation

$\Delta q=1.2\times10^{-6}$ for Recent populations of Sagami Bay

2)   In the case of recessive mutant gene [R-type individuals to be recessive homozygotes only]

$s=1.2\times10^{-5}$

$\Delta q=1.1\times10^{-6}$ for fossil populations of Jizôdô Formation

$\Delta q=1.8\times10^{-6}$ for Recent populations of Sagami Bay

The result of the above calculations indicates for every case that a slight difference of adaptive value (only of the order of 10^{-5} on the average) between the individuals of the two phenotypes would explain the phyletic change of phenotypic frequency. Because their morphological difference is quite significant, I still maintain that natural selection is a probable cause for the observed phenotypic substitution. However, a slight difference of adaptability like this, if present, would not be practically testable with capture—release experiments in the field, because a sample composed of more than a million living individuals would be required for this test.

At the same time I would by no means deny the possibility of non-adaptive causes for this evolutionary change. For example, this slow rate of substitution may also be explicable by spontaneous mutation pressure (only of the order of 10^{-6} per one generation). The effect of random genetic drift, which is most certainly dependent on population size, is also a possible cause, for it is questionable whether such weak selection pressure or mutation pressure could concur with this effect. Regrettably, it must be concluded that the cause of this phenotypic substitution is almost beyond resolution. The current debate between the selectionists and the neutralists would still remain unresolved, even if more ideal fossil records were investigated.

Raup (1977), citing Hayami and Ozawa's (1975) data on the historical change of phenotypic frequency in the lineage of this pectinid (the same data as in my 1973 article), is of the opinion that our observed trend is not necessarily attributable to a deterministic cause like natural selection, because the null hypothesis of random walk cannot be rejected with 95 percent confidence. I agree with him that the trend may be caused by various non-adaptive factors, including stochastic ones. However, Feller's theorem, which was applied by Raup (1977) for the test of random walk, takes into account only the differential frequency of plus and minus signs in a time series of morphological change, and various other valid quantitative data such as net amount of change, number of contemporary samples, sample size, standard errors, and time span between samples are not utilized. I dare say, therefore, that the effectiveness of this test for this purpose is rather weak, and reliable evaluation of random walk cannot necessarily be achieved.

CHAPTER 10

Inferred Speciation of *Cryptopecten*

The process of speciation (multiplication of species) and its relation to the rate of morphological change are one of the most interesting and important subjects in evolutionary paleontology. Various theoretical models and valid new concepts have been put forward in recent years, but it may be said that positive approaches do not necessarily keep up with theoretical studies. This is, of course, primarily due to the inevitable incompleteness of fossil records. Nevertheless, models and theories should be tested in various actual taxonomic groups which are well represented both in fossils and present seas.

As Mayr (1954), Knox (1963), Fretter and Graham (1963) and some others have discussed, there is no reason to consider that the speciation of marine animals is fundamentally different in mechanism from that of terrestrial ones. Though the present data on the speciation of *Cryptopecten* are far from ideal, the origin of each species can be provisionally interpreted in accordance with the widely accepted theory of geographic (allopatric) speciation (Mayr, 1942, 1963; etc.).

There are four living species of *Cryptopecten* in the world; namely, *C. bullatus* in the central-west Pacific, *C. nux* in the Indo-Pacific, *C. vesiculosus* in the seas around Japan and *C. phrygium* in the Gulf of Mexico and adjacent seas (see Fig. 2). They are clearly distinct from one another, not only in morphological characters (e. g., size, shell convexity, outline, surface sculpture and number of radial ribs) but also in geographic and bathymetric distribution (see Figs. 2, 3). Dead shells of the first three species were collected together at a few stations around the southern part of the Japanese Islands, but available records of living specimens indicate that their bathymetric ranges are considerably different; the optimum depth seems to be about 200 meters or still larger in *C. bullatus* and *C. phrygium*, about 100 meters in *C. vesiculosus*, and as shallow as 30 to 50 meters in *C. nux*. I have not seen any individual suggestive of a hybrid.

These three Indo-Pacific species seem to have evolved along independent lineages, at least since Middle Pliocene. In addition to the well-recognized stock of *C. vesiculosus*, the persistent lineage of *C. nux* with little morphological change with time is suggested by several Pliocene and Pleistocene samples in south Japan. Furthermore, according to Cox (1930) and Eames and Cox (1956), *C. nux* is a long-lived species from Early Miocene onward in east Africa. It is therefore probable that *C. nux* is the parental species from which some other species of *Cryptopecten* were derived. The lineage of *C. bullatus*, though its fossil records are still poor, is considered to have been established by the end of Middle Pliocene (Fig. 24).

So far as I am aware, no fossil record of *C. phrygium* has been known. It is a solitary Atlantic representative of this genus but seems to be closely related to *C. bullatus* in the Pacific. Not only their morphology but also their bathymetric ranges are so similar that

78

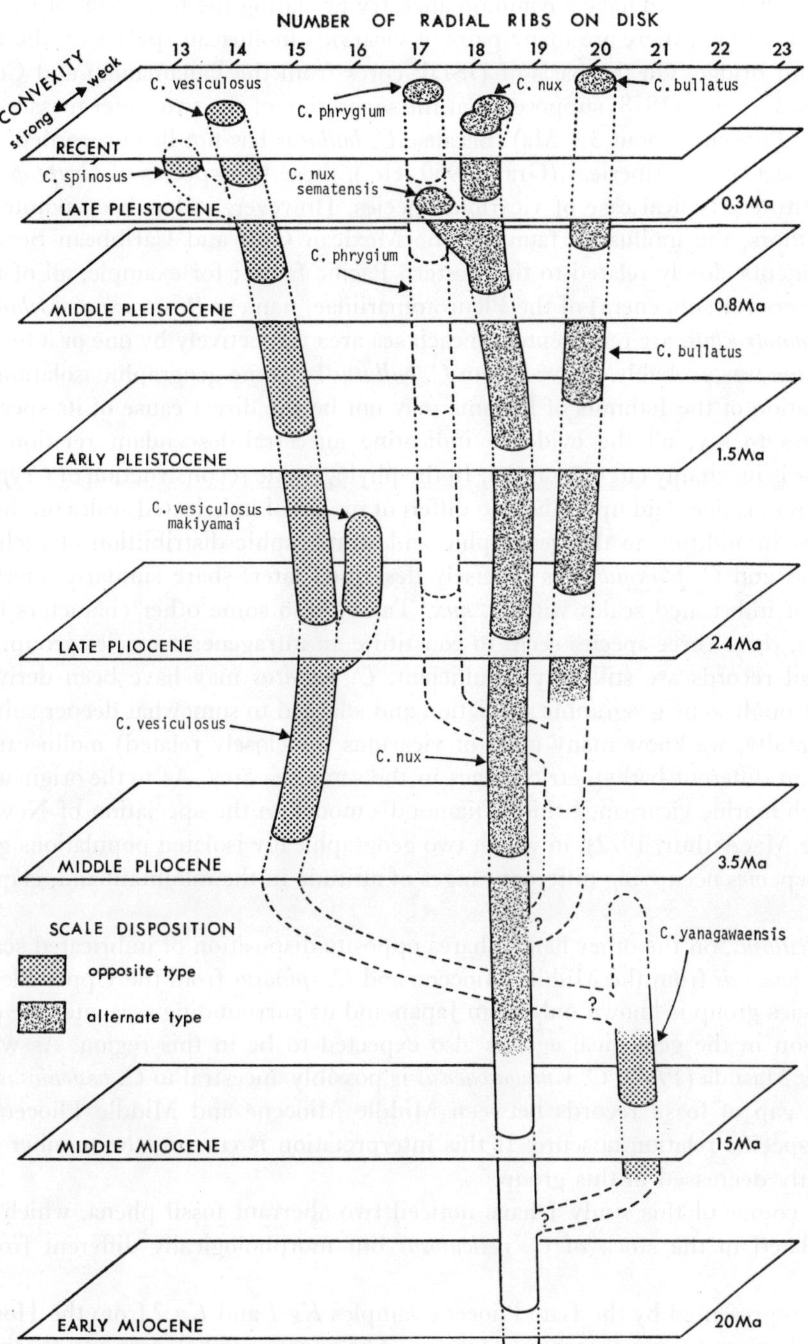

Fig. 24. Inferred interspecific relation of *Cryptopecten* species. Time planes after the Middle Pliocene are spaced in accordance with absolute ages.

the two species seem to have a common ancestry predating the formation of the Isthmus of Panama. In fact, there are many pairs of vicarious molluscan species on the two sides of this land bridge. On the basis of DSDP cores from the Panamanian and Colombian sea basins, Keigwin (1978) supposed that the separation of the two water masses occurred in Middle Pliocene (about 3.1 Ma). Because *C. bullatus* has not been recorded along the western coast of the Americas (Grau, 1959, etc.), these two species of *Cryptopecten* may not constitute a typical case of vicarious species. However, as has been pointed out by many authors, the molluscan fauna of the Mexican Gulf and Caribbean Sea contains some elements closely related to the western Pacific fauna; for example, all of the three living genera (or subgenera) of the Pleurotomariidae, namely *Perotrochus*, *Mikadotrochus* and *Entemnotrochus*, are represented in each sea area respectively by one or a few species. *C. phrygium* was probably derived from *C. bullatus* by some geographic isolation, though the formation of the Isthmus of Panama may not be the direct cause of its speciation.

Needless to say, all the evidence indicating ancestral-descendant relation of fossil organisms is inevitably circumstantial. In the phylogenetic reconstruction of *Cryptopecten*, special stress is here laid upon the two different modes of imbricated scales on the sides of radial ribs in addition to the geographic and stratigraphic distribution of each species. *C. bullatus* and *C. phrygium*, as precisely described later, share similarly alternate disposition of imbricated scales with *C. nux*. Taking also some other characters into consideration, these three species seem to constitute an infrageneric species group. Though their fossil records are still very insufficient, *C. bullatus* may have been derived from *C. nux* through some geographic speciation and adapted to somewhat deeper substratum.

Incidentally, we know many pairs of vicarious (or closely related) molluscan species adapting to different bathymetric ranges in the same sea area. As to the origin and process of such marine vicarism, I think Diamond's model on the speciation of New Guinea birds (see MacArthur, 1972), in which two geographically isolated populations grow into vicariate species occupying different ranges of altitude in the mountainland, is quite suggestive.

C. vesiculosus, on the other hand, shares opposite disposition of imbricated scales with *C. yanagawaensis* from the Middle Miocene and *C. spinosus* from the Upper Pleistocene. This species group is known only from Japan and its surrounding seas, and the center of distribution in the geological ages is also expected to be in this region. As was interpreted by Masuda (1962), *C. yanagawaensis* is possibly ancestral to *C. vesiculosus*, though the wide gap of fossil records between Middle Miocene and Middle Pliocene makes the interspecific relation obscure. If this interpretation is correct, the number of radial ribs greatly decreased in this group.

In the course of this study I have noticed two aberrant fossil phena, which are certainly related to the stock of *C. vesiculosus* but morphologically different from other samples.

One is represented by the Late Pliocene samples *Kg 1* and *Kg 2* from the Hosoya Silt of the Kakegawa Group in central Honshu. The specimens belonging to these samples are characterized by the unusually large ultimate size, rapid growth rate, narrow central solid ridge of each rib, thin test and more numerous radial ribs on the disk (Plate 7, Figs. 11, 12). In the upper part of the Kakegawa Group all the known occurrences of this phenon are from silty or muddy sediments, whereas almost all other fossil samples of

C. vesiculosus came from coarser sediments such as medium- or coarse-grained sands and gravels. At the localities of *Kg 1* and *Kg 2*, the present phenon occurs together with abundant individuals of *Glycymeris rotunda* (Dunker), *Venus* (*Ventricoloidea*) *foveolata* (Sowerby) and *Clementia papyracea* (Gray).

As the result of a number of dredge trials in the coastal area of Sagami Bay, assemblages of molluscs were investigated by Horikoshi (1957, 1960), Biological Laboratory of the Imperial Household (1971) and myself, and it was found that *G. rotunda* and *V.* (*V.*) *foveolata* live together at many muddy bottom stations, but that *C. vesiculosus* is restricted to sandy substrate and is only rarely accompanied by the other two species. At the type locality of the Jizôdô Formation, *C. vesiculosus* and *G. rotunda* occur in adjacent but clearly different fossil beds; the pectinid is abundantly contained in the coarse-grained sandy bed with another glycymeridid, *Glycymeris* (*Tucetilla*) *pilsbryi* (Yokoyama), but never found in the superjacent silty bed where *G. rotunda* is very common. Therefore, the samples *Kg 1* and *Kg 2* are unique not only in terms of morphological characters but also with respect to inferable ecologic habit. At present I regard these two samples as representing populations differentiated from the main stock of *C. vesiculosus* through adaptation to relatively fine-grained substrates. They are provisionally treated here as a new subspecies, *C. vesiculosus makiyamai*, though it is not impossible that this phenon is an incipient species, or merely a product of ecological variation.

The other aberrant fossil form is represented by the sample *Kk (S)* from the Wan Formation of the Ryukyu Group at Kamikatetsu of Kikai Island. This sample is morphologically distinguishable from any sample of *C. vesiculosus* in the much weaker convexity of the two valves, fewer radial ribs, more prominent and granular imbricated scales on the ribs, and several other characters (Plate 8, Figs. 1–4). This phenon occurs dominantly in coral sand in association with *C. nux* and numerous other molluscs (Nomura and Zinbo, 1934), larger foraminifers, calcareous algae, etc.

In addition to these marked differences in the morphology and inferred ecologic habit, this form is regarded as specifically distinct from *C. vesiculosus* for the following reason. According to Sakanoue et al. (1967) and Omura (1983), and the age of this fossil bed is Late Pleistocene (about 0.08 Ma). All the Late Pleistocene and later samples of *C. vesiculosus* in various areas of Japan are, as noted before, composed of the two discrete phenotypes Q and R, while the sample *Kk (S)* is decidedly monomorphic and consists entirely of the individuals comparable with the Phenotype Q in spite of its young age and large sample size. Consequently, the reproductive isolation of this phenon from *C. vesiculosus* was probably completed before the spread of the phenotype R. In this article I describe this peculiar form as *C. spinosus* sp. nov.

Judging from its absence in Recent seas, *C. spinosus* was probably a dead-end branch derived from the main stock of *C. vesiculosus* and was unable to replace this parental species. Because *C. spinosus* is now represented by a large but solitary fossil sample, it is not known whether this was actually a species endemic to the Ryukyu Islands or whether the present sample represents only an immigrant population of a species widely distributed in tropical regions. Further research of Quaternary faunas in southeastern Asia may be necessary in order to solve this problem.

The Late Pleistocene sample *Sm (N)* seems to represent an aberrant phenon of *C. nux* stock, and is characterized by much stronger convexity of the two valves and less developed spiny scales on the byssal wing than is found in other fossil and Recent samples of *C. nux*. This sample is here referred to as *C. nux sematensis*. *C. nux nux* does not occur at the type locality of this subspecies. Since the center of distribution of *C. nux* is believed to be in the equatorial Pacific rather than the Japanese waters, this subspecies probably represents a peripheral population near the northern limit of distribution.

The specimens of *C. nux* from the Red Sea (USNM 764165) seem to be characterized, on the contrary, by unusually tall, weakly inflated and equiconvex shell in comparison with ordinary samples from the Indo-Pacific. Though a subspecific name is not proposed here, the Red Sea population may also represent a peripheral isolate which is now undergoing speciation. It is also interesting to see that the samples of this species from Queensland, also a peripheral area of distribution, show somewhat specialized morphology (see Plate 9, Figs. 2, 5).

In conclusion, the present data of *Cryptopecten* are still too insufficient to test the model of allopatric speciation. It may be said, however, that various facts about the morphology and distribution can be explained to a considerable extent by this theory. At the same time, it is supposed that the short-ranging taxa newly recognized in this study represent only a minor fraction of the speciation and peripheral isolation that have actually occurred.

CHAPTER 11

Discussions on Some General Problems Regarding Patterns of Morphological Change

A longstanding issue in evolutionary paleontology has been how to interpret the morphological gaps between various lineages in the fossil records. Apart from Cuvierian catastrophism and de Vriesian saltationism, the following four views are held, which are neither inconsistent with one another, nor conflict with the accepted mechanisms of microevolution.

1) Superficial morphological discontinuity due to the gap of sedimentation.—This represents the traditional reasoning of stratigraphers and paleontologists, and is said by some punctuationists to have been derived from the notion of "phyletic gradualism". Brinkmann's (1929) interpretation on the morphological break in a lineage of *Kosmoceras* (recently restudied by Raup and Crick (1981, 1982) from the statistical viewpoint) and Newell's (1956, fig. 3) generalized diagram, in which a lineage with continuous and unidirectional change of morphology is interrupted by some gaps of sedimentation, are typical examples of this view.

2) Essential morphological discontinuity due to the inconstant rate of morphological change within a lineage.—If stepwise morphological change occurs within a lineage, it is assumed that evolutionary patterns generally become intermittent, since the process of rapid transformation can rarely be traced in the fossil records. This view was popularized by Simpson (1953), who considered that the origin of a higher taxon can often be sought in some rapid breakthrough of a threshold between different adaptive zones.

3) Essential morphological discontinuity due to allopatric speciation and subsequent migration.—This view is an expansion of Mayr's (1942, 1963) theory of geographic speciation—that is, a new species arises rapidly from a peripherally isolated small population. This model was, I think, substantially adopted by Simpson (1953) in his interpretation of the trilobite sequence in the Cambrian of Sweden, and recently much emphasized by Eldredge and Gould (1972) and others as an alternative to phyletic gradualism. In this model it is generally anticipated that rapid genetic change at speciation results in rapid morphological change and that ecologic replacement of an existing species by an immigrant new species results in a punctuated pattern in the fossil records.

4) Essential and intrinsic morphological discontinuity due to phenotypic substitution.—If a single or a few genetic factors strongly influence morphological difference in an evolving population, phenotypic substitution (or the process from the introduction to the fixation of a mutant) would result in essential morphological discontinuity (Hayami and Ozawa, 1975). This process is quite different from the saltationists' "systemic mutation" and "Grossmutation", but, whether adaptive or non-adaptive, is not inconsistent with the acknowledged mechanism of evolution.

Morphological discontinuity in the fossil records, I believe, may occur due to various geological and biological causes such as these. Among others, sedimentary gaps may be ubiquitous geological phenomena, representing a quite extrinsic cause for morphological discontinuity of various scales, because they must occur regardless of evolutionary pattern. This notion is therefore, by no means the exclusive property of the gradualists. Pure gradualism may be better defined in this case as the notion that all the observed morphological gaps should be attributed to the incompleteness of fossil records. The three other possible causes are biologically more fundamental. The problem is their relative frequency and scale rather than their adequacy.

As was already discussed in the preceding chapter, the evolution of *Cryptopecten* seems to be characterized by the presence of a few well-established, long-ranging species and the sudden appearance of some short-ranging unsuccessful branches. Some phyletic (sometimes directional) change actually occurred in the lineage of *C. vesiculosus* and possibly also in other lineages, but the relatively conservative morphology of long-ranging species and the rareness of morphologically transitional fossil samples may be consistent with the punctuational model.

The proposal of the punctuational model of morphological change (Eldredge and Gould, 1972; Gould and Eldredge, 1977) had enormous impact on evolutionists and has recently developed into more comprehensive theories of macroevolution such as higher-level selection and hierarchical concept of evolution (Stanley, 1979; Gould, 1982; etc.). As emphasized by some of these authors, this model, though potentially testable, has not yet been tested to satisfaction. To do so would require excellent material and patient field and laboratory work. Immediately after the first proposal of this model, many paleontologists attempted to test it on the basis of their own data; some were found to agree but others not. So far as I am aware, however, most of the evidence cited to support this model is still circumstantial. On the other hand, gradual change of morphological unit characters was demonstrated in some lineages of animals by recent morphometric studies at the population level. Though well-documented examples of long-ranging, gradual transformation seem to be restricted to some protozoan lineages (Ozawa, 1975; Kellogg, 1975; Malmgren and Kennett, 1981; etc.), it is certain that morphology often changes gradually with or without a definite trend in metazoan lineages. Phyletic size increase is common in various lineages of Jurassic Bivalvia in western Europe (Hallam, 1975). Phyletic decreases in shell size and number of radial ribs are well recognized in the stock of *C. vesiculosus*, offering another example of gradual change within a lineage.

Nevertheless, I am unable to agree with the theory of phyletic gradualism. The actual macroscopic pattern of morphological change must, I think, resemble the punctuational model. Paleontologists are want to give a decision in favor of the party possessing the more plentiful circumstantial evidence. In this case, however, the debate between the gradualists and the punctuationists should not be decided so easily, for since geographic (allopatric) speciation, as considered by Mayr, probably occurs in a geological instant, and the process is bound to be less frequently recorded in fossil sequences than that of phyletic evolution.

Very recently, Williamson (1981) reported an interesting sequence of molluscan faunas in Late Cenozoic lacustrine deposits in the Turkana Basin of northern Kenya, discussing this current problem on the basis of actual paleontological data. Though the full account would be desirably published in some separate article, his material probably represents the best documented process of molluscan speciation that we have. According to him, in the lacustrine deposits, there are about ten long-ranging lineages of gastropods and bivalves with various reproductive strategies, which are morphologically very conservative throughout the sequence. At two definite horizons, however, new forms (possibly new species) arose simultaneously through rapid morphological change from many of these stocks, but they survived only temporarily to be suddenly replaced by the populations of conservative morphology without any intermediates. These incipient forms, which were said to have arisen from isolated populations by lacustrine regressions, seem to have become extinct when their parental species came back to the same lake through succeeding transgressions. Williamson interpreted these evolutionary patterns as being in evident conformity with Mayr's "founder principle" as well as with Eldredge and Gould's "punctuated equilibria". However, it is generally difficult to reconstruct the process of speciation in the case of marine molluscs from the data in this kind of local sequence.

My own experience leads me to feel that the theory of punctuated evolution is valid. Many morphological gaps may be caused by such episodic events as speciation, migration and extinction. I feel that Gould and Eldredge's (1977) claim, "Stasis is data", deserves serious consideration by evolutionary paleontologists. In current evolutionary studies, as they pointed out, the dynamic nature of a lineage is apt to be unfairly emphasized at the expense of the static. Thus although I concluded in the present study the directional shift of three characters, i.e., shell size, average number of radial ribs, and relative frequency of mutant phenotype, in the lineage of *C. vesiculosus*, but significant evolutionary change could not be detected for many other morphometric characters. It is tentatively supposed that these "static" characters are greatly influenced, not only by genetic causes, but also by environmental factors, and that the wide range of variation (or the too small sample size) makes it difficult to recognize chronological change. At the same time, however, I cannot necessarily deny the possibility that they are truly static.

Aside from the discrete variation of sculpture in *C. vesiculosus*, the net amount of observed morphological change within each lineage of *Cryptopecten* is, in every way, much less significant in comparison with the morphological difference between distinct lineages. The results of this study also suggest the possibility that many taxa (either species or subspecies) may arise from well-established stocks through isolation. Some of these isolated populations may grow into distinct species, and others may become extinct or be absorbed in the parental population in the main distribution. I believe that the three short-ranging taxa of *Cryptopecten*, i. e., *C. vesiculosus makiyamai*, *C. nux sematensis* and *C. spinosus*, represent only a fraction of the isolation and speciation that actually occurred. Speciation may be a ubiquitous event, many other unsuccessful, dead-end branches having been lost from the fossil records. This is why, in spite of the recognized gradual morphological change in the lineage of *C. vesiculosus*, I prefer the punctuated model.

A few reservations remain, however. Firstly, the current punctuationists' reason for attributing the cause of rapid morphological change almost exclusively to speciation still fails to completely convince. May not rapid morphological change also occur through the breakthrough of adaptive threshold or through bottle-neck effect within a single lineage? Secondly, the view that reproductive isolation is always accompanied by significant morphological shift (or *vice versa*) may be too straightforward, given the many phenomena in nature (e. g., sibling species and polytypic species) where this is not the case.

Finally, I will discuss the significance of phenotypic substitution in the pattern and process of evolution. Though well-documented examples are still scarce, this is the most intrinsic cause of punctuated pattern. Frequency is not a problem, because polymorphism is a widespread phenomenon in living organisms. Questions about its role in macroevolution concentrate on scale:—how can significant morphological change be accomplished by a single mutation. Gould and Eldredge (1977) regarded my *Cryptopecten* data of phenotypic substitution (Hayami, 1973) as being too microscopic to test their punctuational model. Certainly, phenotypic substitution is not a case of transspecific evolution but represents a peculiar example of phyletic evolution.

It has been generally supposed that a new species or a higher taxon cannot arise from a single gene or chromosomal mutation. Recent advancement in the studies of molecular biology, however, seems to suggest that significant evolutionary change may occur within a lineage by the mutation of a "regulatory gene" and its fixation in the population. With the diffusion of modern synthetic evolutionary theory, Goldschmidt's (1940) anticipation of a "hopeful monster" was nearly abandoned for its too extreme notion of the instantaneous establishment of a new higher taxon. Nevertheless, polyploidy is still credible as a possible cause of instantaneous speciation. According to Menzel (1968) and Patterson (1969), chromosome numbers are relatively stable among the species within a bivalve family, but considerably variable among different families. Though cytological studies of bivalves seem to lag far much behind those of gastropods (especially non-marine species), the role of polyploidy and other chromosomal change in macroevolution (especially the punctuated pattern of morphological change) is an interesting theme in evolutionary paleontology and should be investigated. Moreover, as was discussed by Stanley (1979) and others, future studies of "regulatory genes" may result in the resurrection of the macrogenesis theory, though in a form quite different from Goldschmidt's concept.

The observed morphological gap between the two phenotypes of *C. vesiculosus* is so clear that I wonder why few previous taxonomists regarded them as specifically distinct. If only the features of surface sculpture were taken into consideration, the difference between the two phenotypes would appear to be more significant than that between different species. I also expect that some related but morphologically distinct nominal molluscan "taxa" will prove to constitute discontinuous variation within an interbreeding population. The relation between *Volachlamys hirasei* and *V. awajiensis* may prove to be such a case (see p. 59). Evaluation of the role of phenotypic substitution depends upon careful studies of the relation between phena and taxa in living and fossil populations.

As was diagramatically shown in Fig. 24 and will be described in the chapter on systematics, in *Cryptopecten* there are two alternative modes in the disposition of imbricated scales on the sides of radial ribs. One is the alternate type, as seen in all the observed individuals of *C. bullatus* (Plate 2, Fig. 3e: Plate 11, Figs. 3a, b), *C. nux* (Plate 2, Fig. 4g) and *C. phrygium* (Plate 9, Figs. 6a, b, 7a, b). The other is the opposite type, as seen in the fresh specimens of *C. yanagawaensis* (Masuda, 1958, pl. 27b, fig. 8), *C. vesiculosus* (Plate 5, Figs. 1e, f, 2c, d, 3e, f; Plate 11, Figs. 1a, b, 2a, b) and *C. spinosus* (Plate 8, Fig. 5; Plate 12, Fig. 5). Because the relation between the two species groups has not yet been clarified on the basis of concrete evidence, it is now difficult to say which type is the more primitive. Aside from the phylogenetic relation, gradual dislocation of scales from one type to the other is rather unlikely, because the periodicity is regular in both types. The ultimate origin of this change, therefore, should be sought in a single mutation; that is to say, one type may have arisen from the other through phenotypic substitution. If the phenotypic substitution were rapid, the process could hardly be expected to be recognizable in the fossil records. It would likely appear to be a case of an ancestral "species" suddenly becoming extinct and a descendant "species" suddenly appearing, as was discussed by Hayami and Ozawa (1975). Species and higher taxa are often discriminated on the basis of alternative character, which may be possibly switched by a rather simple genetic cord. If the suggested existence of "regulatory genes" is taken into consideration, the role of phenotypic substitution occurring in a geological instant, particularly as the ultimate origin of a new alternative taxonomic character, should not be underestimated in the studies of macroevolution.

CHAPTER   12

Systematic Description of *Cryptopecten*

Class Bivalvia Linnaeus, 1758
Subclass Autobranchia Grobben, 1894*
Superorder Pteriomorphia Beurlen, 1944*
Order Ostreoida Férussac, 1822*
Suborder Pectinina Waller, 1978*
Superfamily Pectinacea Rafinesque, 1815
Family Pectinidae Rafinesque, 1815
Subfamily Chlamydinae Korobkov, 1960

Genus *Cryptopecten* Dall, Bartsch and Rehder, 1938

1938. *Cryptopecten* Dall, Bartsch and Rehder, *Bishop Mus. Bull.*, vol. 153, p. 93.
1939. *Corymbichlamys* Iredale, *Brit. Mus. (Nat. Hist.), Sci. Rept.*, vol. 5 (Mollusca, pt. 1), p. 367.
1951. *Cryptopecten* Dall, Bartsch and Rehder: Habe, Genera of Japanese Shells, p. 77.
1969. *Cryptopecten* Dall, Bartsch and Rehder: Hertlein *in* Cox et al., Treatise on Invertebrate
Paleontology, pt. N, vol. 1, p. N357. [as a subgenus of *Chlamys*]
1973. *Cryptopecten* Dall, Bartsch and Rehder: Fatton, *CERPAB, Notes et Contr.*, no. 8, p. 46.
[as a subgenus of *Chlamys*]
1977. *Cryptopecten* Dall, Bartsch and Rehder: Habe, Systematics of Mollusca in Japan, p. 84.

Type-species.—Cryptopecten alli Dall, Bartsch and Rehder, 1938 [=*Pecten (Chlamys)*
bullatus Dautzenberg and Bavay, 1912]; Hawaii, Philippines and Japan, Late Pliocene to
Recent.

Historical review.—The generic name *Cryptopecten* was proposed by Dall, Bartsch and
Rehder (1938), who designated *C. alli* from the sea around Hawaii as the type-species.
The original diagnosis runs as follows:

"Shell of medium size, orbicular, laterally compressed, moderately thin; wings
subequal, byssal sinus conspicuous, with a few short lamellar denticles on the lower
border. Sculpture consists of radiating ribs which are very regularly lamellose, the
laminations extending ventrally where they are met by the next lamination with which
they fuse. They thus enclose a series of hollow chambers."

The characteristic sculpture of *Cryptopecten* was partly expressed in the second half of
the original diagnosis. Dall, Bartsch and Rehder (1938) probably regarded some other
species as belonging to *Cryptopecten*, because they wrote, "This group appears to have
an Indo-Pacific distribution, for we have seen specimens from the Philippines and Fiji,
as well as Australia". However, they neither indicated any other specific name as other
members of *Cryptopecten*, nor did they compare it with any other genus.

On the other hand, Iredale (1939) proposed genus *Corymbichlamys*, designating

---

*   The suprafamilial classification by Waller (1978) is here adopted.

88

Chlamys corymbiatus Hedley, 1909, from northeastern Australia as its type-species. Hertlein *in* Cox et al. (1969) regarded *Corymbichlamys* as congeneric with *Argopecten*, but I think that it should be treated as a junior synonym of *Cryptopecten*, because *C. corymbiatus* is synonymous with *Pecten nux* Reeve, 1865.

Subsequently, Habe (1951, 1977) recognized *Cryptopecten* as a distinct genus including such Japanese species as *Pecten vesiculosus* Dunker, 1877, "*Pecten tissotii* Bernardi, 1858" [=*C. bullatus*], *Pecten nux* Reeve, 1865, *Pecten inaequivalvis* Sowerby, 1887, and *Pecten oweni* Gregorio, 1936. The generic diagnosis given by him is somewhat different from the original and stresses the stronger convexity of the right valve and the larger size of the anterior wing. Many Japanese malacologists and paleontologists followed him in these generic assignments, but some others did not. For example, Taki and Oyama (1954) and Oyama (1973) regarded *Cryptopecten* as a subgenus of *Aequipecten* Fisher, 1886, and Masuda (1962) and Masuda and Noda (1976) referred *P. vesiculosus* and some related fossil species to *Aequipecten*.

Outside Japan, Hertlein *in* Cox et al. (1969) treated *Cryptopecten* as a subgenus of *Chlamys*, and regarded *Gloripallium* Iredale, 1939, as a junior synonym of *Cryptopecten*. He gave the following diagnosis to this subgenus, "Major ribs and interspaces sculptured with radial riblets covered with fluted scales; cardinal crura present". Fatton (1973) also regarded *Cryptopecten* as a subgenus of *Chlamys*. Kay (1979) did not accept *Cryptopecten* as a genus-group taxon, but regarded its type-species, *C. alli*, as belonging to *Chlamys*. Thus many different opinions have been presented about the validity and application of the generic (or subgeneric) name of *Cryptopecten* as well as its limits and taxonomic position.

Generally speaking, the classification of Cenozoic and Recent Pectinidae is still in confusion. Pectinids include many conspicuous species which are attractively shaped and beautifully colored, and their taxonomic description began in the later half of the 18th century. Because of insufficient description and poor locality information, the nature of many early described species remains obscure. As listed by Vokes (1967, 1980) and Hertlein *in* Cox et al. (1969), more than 120 taxa of genus-group have been proposed, but their diagnostic characters and limits as well as relationship to other taxa are not necessarily clear. The same holds true for the subfamilial division of this family. For instance, many genera characterized by deep byssal notch, well-developed ctenolium, relatively tall outline, scarcely gaped disk margin and scaly radial ribs may be grouped in the subfamily Chlamydinae Korobkov, 1960, but their limit and relationship with other subfamilies (e. g., Pectininae) are still very much debatable. In order to settle the classification system of this family, more comprehensive studies should be done on the basis of the materials that exist worldwide and on sound evaluation of taxonomic characters. It goes without saying that studies of fossils are important for the recognition of phylogenetic relationship.

In this article discussion is confined to the validity and morphological characteristics of *Cryptopecten*, but it seems necessary to first clarify its discrimination from other related genera of the Chlamydinae.

Among the living Pacific pectinids referred to *Cryptopecten*, *Pecten vesiculosus* and *Pecten nux* share many essential characteristics with the type-species, *C. alli* [=*Pecten (Chlamys) bullatus*]. On the other hand, *Pecten inaequivalvis* is similar to the type-species

90

of *Haumea*, i. e. *H. juddi* Dall, Bartsch and Rehder, 1938 [=*Pecten loxoides* Sowerby, 1882], and should be excluded from the present genus. Specific characters of *Pecten oweni* are not very clear at present, but specimens in Japan hitherto referred to this species (e. g., Kuroda, 1932; Shikama, 1964) do not have any features of the characteristic *Cryptopecten* sculpture. The radially striated ribs of *P. oweni*, so far as these Japanese specimens are concerned, remind me, in fact, of those of *Comptopallium* Iredale, 1939, and *Decatopecten* Rüppel *in* Sowerby, 1839.

Pecten phrygium Dall, 1886, from the Mexican Gulf is, as was pointed out by Waller (personal communication, April 21, 1981) and treated by Woodring (1982), undoubtedly another representative of *Cryptopecten*, because the surface sculpture and many other characters are essentially similar to the type-species. *Pecten muscosus* Wood, 1828, from the Mexican Gulf, which was assigned to *Cryptopecten* by Shikama (1964), does not seem to belong to this genus in view of the large wings and spiny (not imbricated) scales on the radial ribs. *P. muscosus*, I maintain, constitutes another species group together with *Pecten acanthodes* Dall, 1925, both of which were referred to *Aequipecten* by Waller (1969, 1973). Ladd (1945) regarded *Pecten historionicus* Gmelin, 1791, as a doubtful member of *Cryptopecten*, but this species, as was clarified by Waller (1972a), is an undoubted species of *Excellichlamys*.

Recently Woodring (1982) redescribed *Pecten cactaceus* Dall, 1898, from the Upper Miocene (or Lower Pliocene) of Panama, regarding it as an early representative of *Cryptopecten*. As he pointed out, the sculpture of this species is very unique, being characterized by several persistent intercostal threads covered by fine imbricated scales. The hollow structure (see Woodring, 1982, pl. 124, figs. 9, 11) strongly reminds me of that of Phenotype *R* of *Cryptopecten vesiculosus*. As a result of my observation of the Panamanian specimens in the National Museum of Natural History, Washington, D. C., however, *Pecten cactaceus* seems to be considerably different from *C. vesiculosus* and other representatives of *Cryptopecten* because of the undeveloped ctenolium, well-developed and persistent intercostal threads, relatively shallow byssal notch, and unusually thin shells. Its generic reference to *Cryptopecten* is therefore still debatable. At present, it may be adequate that an emended diagnosis of *Cryptopecten* considers common characteristics of *P. vesiculosus*, *P. nux* and *P. phrygium* in addition to the type-species (*P. bullatus*) and a few undoubted fossil species.

Emended diagnosis of Cryptopecten.—Shell comparatively small-sized for pectinids, with height subequal to length, nearly acline but tending to become slightly prosocline in later growth stages. Convexity of shell variable among species, but right valve commonly more strongly inflated than left, the reverse being the case in early growth stages. Byssal notch moderately deep, provided with several denticles of ctenolium. Wings moderate in size; anterior generally larger than posterior. Anterodorsal margin nearly straight, posterodorsal margin slightly concave, especially in later stages. Apical angle moderately large for pectinids. Disk margin scarcely gaped. Disk suborbicular, ornamented with 12 to 25 strong and simple radial ribs, and in later growth stages with a few fine threads on each interspace as well. Both lateral sides of radial ribs, and sometimes interspaces, covered with fine imbricated scales which enclose narrow hollow chambers. Wings of both valves with several radial ribs without hollow chambers. Early dissoconch marked with delicate

Camptonectes-like striae, occurring from much earlier stage in left valve than right and disappearing before shell attains 4 mm in height. Coloration quite variable within each species, but commonly reddish brown. Left valve more darkly pigmented than right. Byssal wing of right valve almost invariably pale. Outer ligament area comparatively thin. Resilial pit small or moderate in size.

Comparisons.—One of the striking characteristics of *Cryptopecten*, as stressed in the original description, lies in the mode of radial ribs, which is characterized by fine imbricated scales covering the sides of the ribs with some hollow space. The radial ribs are quite regular in prominence, neither bifurcated nor inserted except for very early stages. Another conspicuous feature of this genus is the highly allometric growth of shell, which is quite different in the two valves; the shell turns from left-convex to right-convex through equiconvex. These features may not be exclusive, but they are also sometimes observable in certain species of other genera of the Chlamydinae. The taxonomic distinction and validity of *Cryptopecten* at the generic level, however, are supported by the combination of these characteristics.

Some Japanese authors, as noted above, treated *Cryptopecten* as a subgenus of *Aequipecten* Fischer, 1886 [type-species: *Ostrea opercularis* Linnaeus, 1758, monotypy]. The name of *Aequipecten* seems to have been widely and sometimes inadequately used by many workers for fossil and living pectinids. The type-species of *Aequipecten* has left-convex shell throughout growth, almost solid radial ribs and widely gaped anterodorsal and posterodorsal margins. In England *Aequipecten opercularis* is called "Queen scallop", and its anatomical features and mode of life are known in detail (Dakin, 1909; Rees *in* Cox, 1957). The wide gapes of disk are probably related to swimming habit (Stanley, 1970). On the contrary, *Cryptopecten* is, so far as I observed in *C. vesiculosus*, not an active swimmer, and the disk margins are scarcely gaped. Though the suborbicular disk, unequal wings, and regular and simple (neither bifurcated nor inserted) radial ribs of *Cryptopecten* are similar to those of *Aequipecten*, the two are quite unlike in the marginal gaping and mode of relative growth. So far as I am aware, there is no true living representative of *Aequipecten* in the Indo-Pacific.

The diagnostic characters and limits of *Aequipecten* are, however, not necessarily clear. As the result of my preliminary survey, it may be classified into the following three species groups:

1) Group of *Aequipecten opercularis* (Linnaeus, 1758) mainly from the eastern Atlantic, characterized by equiconvex or slightly left-convex shell and wide gape of disk margin.

2) Group of *Aequipecten muscosus* (Wood, 1828) from the northwestern Atlantic and Mexican Gulf region, characterized by equiconvex shell, closed disk margin, large wings and common occurrence of spiny scales on radial ribs.

3) Group of *Aequipecten flabellus* (Gmelin, 1791) from the eastern Atlantic and Mediterranean, characterized by equiconvex or right-convex shell, narrow gape of disk margin and relatively small wings.

The first group is, of course, typical *Aequipecten*. The second group including *Aequipecten acanthodes* (Dall, 1925), I think, may eventually be generically or subgenerically separable from the first group. The third group includes such right-convex species as

Fig. 25. Sculpture dimorphism of *Aequipecten commutatus* (Monterosato), ×3. The two specimens belong to one and the same sample (USNM 764279) from the western Mediterranean (off Melilla, 20 fms). Left: a left valve with erect scales on radial ribs. Right: a left valve with imbricated scales covering a pair of narrow hollow parts on both sides of radial ribs.

Aequipecten commutatus (Monterosato, 1875) from the Mediterranean and west coast of north Africa and *Aequipecten atlanticus* (Smith, 1890) from the sea around St. Helena Island.

A. commutatus is a particularly impressive species, because it often displays a hollow structure of radial ribs somewhat similar to *Cryptopecten*, and because some dimorphism is observed in the surface sculpture (Fig. 25). Among several samples of this species in the National Museum of Natural History which I have examined, the sample USNM 764279 from Morocco (off Melilla, 20 fms) is the largest, consisting of 35 complete conjoined valves. In 28 individuals of this sample, each flat-topped radial rib is accompanied by a pair of narrow hollow parts which are covered with fine imbricated scales. On the other hand, hollow structure is completely absent at all in the seven remaining individuals. Although this dimorphic phenomenon should be studied on the basis of more samples and field observation, this may provide a clue to the origin of hollow structure. At the least, it seems to suggest that hollow structure has spread in the populations of *A. commutatus* through phenotypic substitution. In many respects *A. commutatus* and its allies appear to be intermediate between typical *Aequipecten* and *Cryptopecten*.

Hertlein *in* Cox et al. (1969) regarded *Gloripallium* Iredale, 1939 [type-species: *Ostrea pallium* Linnaeus, 1758, original designation] as a junior synonym of *Cryptopecten*. The tripartite radial ribs of *G. pallium* may be apparently similar, but the scales on the ribs and interspaces are much coarser and are never imbricated as in the species of *Cryptopecten*. The ctenolium in young growth stages is exposed on the external surface along the boundary between the byssal wing and disk in every species of *Cryptopecten*, but is completely concealed by the byssal wing in *Gloripallium*. Moreover, the disk of *Gloripallium* is nearly equiconvex, nearly acline throughout growth and similarly pigmented in the two valves. As was pointed out by Waller (1972a), some internal thickening of the wings is observed in *Gloripallium*, but not in *Cryptopecten*. *Gloripallium* and *Cryptopecten* should be regarded as distinct.

Argopecten Monterosato, 1889* [type-species: *Pecten solidulus* Reeve, 1853 (=*Pecten circularis* Sowerby, 1835), subsequent designation by Monterosato, 1899] may also be a genus comparable with *Cryptopecten*. This genus is well represented on the Atlantic coast of the United States, Gulf of Mexico and Caribbean Sea, and a few species are also known in the eastern Pacific from California to Chile (Grau, 1959; Waller, 1969). Fossil *Argopecten* from the Neogene and Quaternary of North America were thoroughly studied by Waller (1969), and offered an outstanding example of molluscan evolution. The shell outline and convexity of *Argopecten* are considerably variable, but, as described by Waller (1969) and others, the shell is generally equiconvex and nearly acline throughout growth, and the wings are subequal in size. Consequently, the posterior wing is commonly much larger than that of *Cryptopecten*. In *Argopecten*, so far as I am aware, the radial ribs are solid without any hollow structure, and scales are generally undeveloped on the ribs and interspaces. *Plagioctenium* Dall, 1898 [type-species: *Pecten ventricosus* Sowerby, 1842 (=*Pecten circularis* Sowerby, 1835), original designation] is regarded as synonymous with *Argopecten*.

Haumea Dall, Bartsch and Rehder, 1938 [type-species: *Haumea juddi* Dall, Bartsch and Rehder, 1938 (=*Pecten loxoides* Sowerby, 1882), original designation] from the shallow waters around the Hawaiian Islands may be related to *Argopecten* in view of the smoothish radial ribs and interspaces, and undeveloped ctenolium. The right-convex shell throughout growth, very small anterior wing, shallow byssal notch, and decidedly prosocline disk, however, seem to qualify it for designation as a distinct genus. As noted before, *Pecten inaequivalvis* Sowerby, 1887, from east Asia seems to belong to this genus. Whatever the case, *Cryptopecten* appears to be unrelated to *Haumea*.

Annachlamys Iredale, 1939 [type-species: *Pecten leopardus* Reeve, 1853, original designation], mainly from the shallow waters of the western tropical Pacific, clearly differs from *Cryptopecten* in the shallow byssal notch, subequal wings, and rough growth lamellae on the surface. Though weak denticles of ctenolium are observed in young individuals of a Japanese representative, *A. reevei* (Adams and Reeve, 1850), this genus is apparently intermediate between the Chlamydinae and the Pectininae.

Volachlamys Iredale, 1939 [type-species: *Pecten cumingii* Reeve, 1853 (=*Pecten singaporinus* Sowerby, 1842), original designation], from the shallow waters of the western Pacific, is clearly different from *Cryptopecten* in the much larger size, acutely triangular posterior wing, undeveloped scales on the surface, and solid radial ribs. A Japanese representative, *Volachlamys hirasei* (Bavay, 1904), shows highly allometric growth and dimorphic phenomenon comparable with the case of *Cryptopecten vesiculosus*, but these features do not seem to indicate that the two genera are closely related.

---

* There are several quite different opinions about the contents of the genus (or subgenus) *Argopecten*, including the nature of its type-species, *Pecten solidulus* Reeve, 1853. Though I am not in a position to discuss this nomenclatorial problem, I here follow Waller's (1969, pp. 32–34) interpretation, in which the type specimen of *P. solidulus* is referable to *Pecten circularis* Sowerby, 1835, from the eastern Pacific (not to such Mediterranean species as *Pecten commutatus* Monterosato, 1875); and, therefore, *Plagioctenium* Dall, 1898, should be treated as a junior synonym of *Argopecten*.

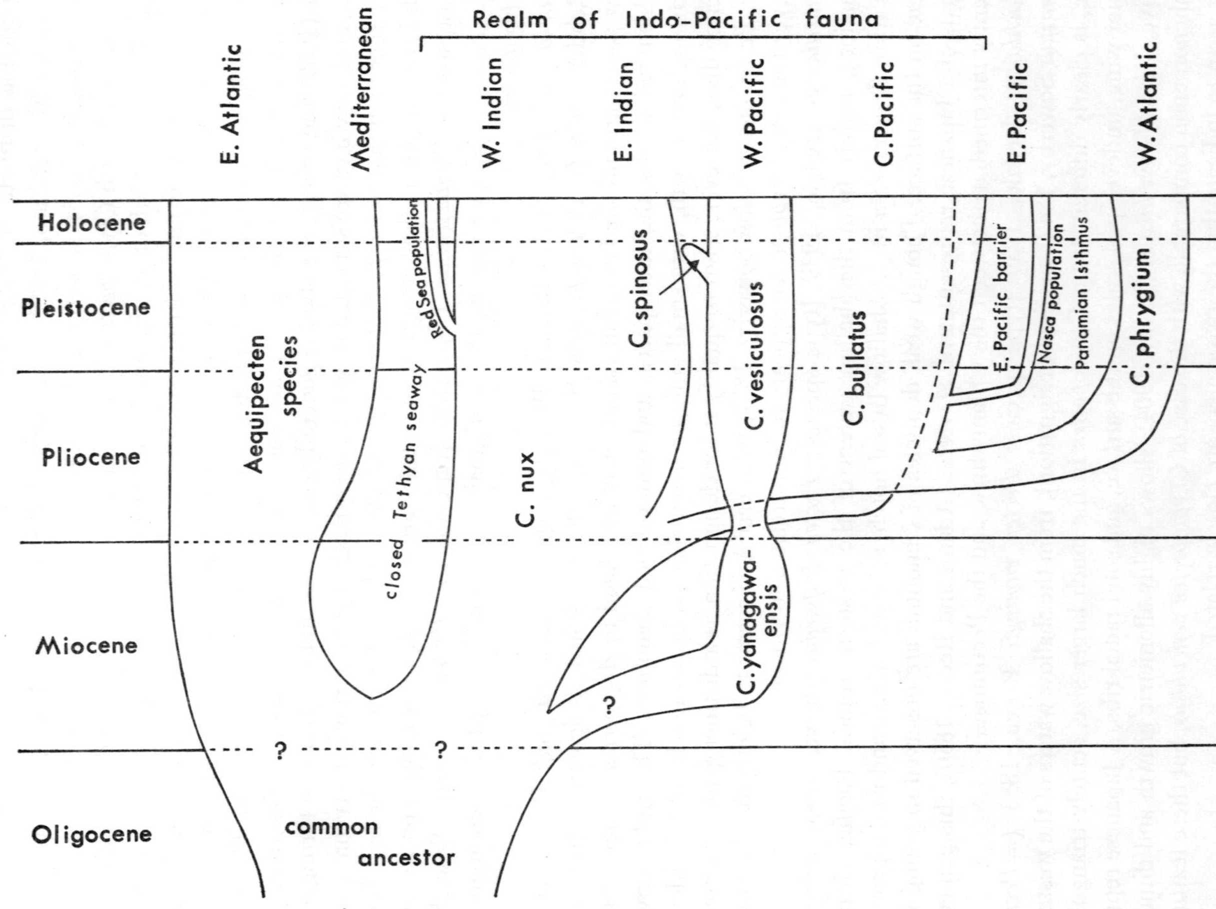

Fig. 26. Generalized diagram of biogeography (longitudinal distribution) of *Cryptopecten* through geologic time. Absolute age and geographic distance are not to scale.

Excellichlamys Iredale, 1939 [type-species: *Pecten spectabilis* Reeve, 1853] from the western Pacific and Indian Ocean may be somewhat related to *Cryptopecten*. As was clearly illustrated by Waller (1972a, pl. 6, figs. 93–96), *E. spectabilis* shows somewhat similarly imbricated scales which cover the radial ribs with some hollow space. Though *E. spectabilis* and *Excellichlamys histrionica* (Gmelin, 1791) from Japan seem to show wide variation in shell outline and surface sculpture, *Excellichlamys* can be generally distinguished from *Cryptopecten* by the fewer and often bifurcated radial ribs, undeveloped central solid ridge of each rib, weaker convexity of left valve, and presence of spotted color pattern on the disk.

The generic validity and distinctness of *Cryptopecten* are thus confirmed. As were collectively described above (p. 22), the surface characters of early dissoconch are nearly the same for the three extant Indo-Pacific species of this genus. Important morphological characters for specific and subspecific discrimination are shell size, convexity of both valves, thickness of test, apical angle, number of radial ribs on disk, size and mode (either opposite or alternate) of imbricated scales covering the sides of radial ribs, shape and sculpture of byssal wing, development of inner shell layer, and presence or absence of dimorphism in surface sculpture.

Origin of Cryptopecten.—Little has been studied about the ancestry of genus *Cryptopecten*. The earliest undoubted representative of this genus so far known is *C. yanagawaensis* from the lower Middle Miocene of Japan in the Pacific and *C. nux* from the Lower Miocene of Tanzania in the Indian Ocean. The two species, I presume, may have constituted independent stocks in different regions during Early-Middle Miocene times, but it is now difficult to determine which stock is parental and what their common ancestor is.

In this connection, however, it is worthy of notice that there are several Oligocene and Early Miocene pectinids which are considerably similar to the known species of *Cryptopecten* in general outline and some other shell characters. Such species are especially common in the Oligocene and Miocene of Iran, east Africa and Mediterranean region. Many of them were described under the subgeneric name of *Aequipecten* by Cox (1927, 1930) and Eames and Cox (1956). "*Aequipecten*" in Eames and Cox' sense appears to be a somewhat heterogeneous group, to which they referred, for example, *Chlamys townsendi* and *Gloripallium pallium*. Nevertheless, some Oligocene and Miocene species from this region may be actually ancestral to the true Recent representatives of *Aequipecten* in the Mediterranean and eastern Atlantic.

It is generally admitted that the marine molluscan fauna of the western Indian Ocean was intimately connected with that of the Mediterranean region until the beginning of Miocene. For example, several species of *Aequipecten* from the Oligocene of Iran have also been known from Palestine, Italy and south France. After Early Miocene the Tethyan seaway was probably interrupted, and the provinciality of the two sea realms seems to have become clear. On the Indian side, *Aequipecten* seems to have declined after Miocene, and probably is not represented by any undoubted species in the present sea.

Chlamys (*Aequipecten*) *deltoidea* Cox, 1927, from the Lower Miocene of Pemba Island, Tanzania, *C.* (*A.*) *iranica* Eames and Cox, 1956, from the Lower Miocene of Iran, and *C.* (*A.*) *kiwazinensis* Eames and Cox, 1956, and *C.* (*A.*) *pseudotjaringenensis* Eames and Cox, 1956, from the Middle Miocene of Iran are rather unlike typical Recent species of *Aequipecten*, but resemble, though probably only superficially, *Cryptopecten yanagawa-*

96

ensis and *C. vesiculosus* in outline and number of radial ribs. However, these Miocene species do not show any distinct hollow structure of radial ribs as is commonly seen in *Cryptopecten*. Further studies of Paleogene and Miocene pectinids in the Tethyan region seem to be necessary in order to determine the phylogenetic relation between *Aequipecten* and *Cryptopecten*.

Cryptopecten bullatus (Dautzenberg and Bavay)

Pl. 1, Figs. 1–6; Pl. 2, Figs. 1–3; Pl. 9, Fig. 1;
Pl. 10, Fig. 3; Pl. 11, Fig. 3

?1882. *Pecten tissoti* [sic] Bernardi: Dunker, Index Molluscorum maris Japonici, p. 241. [non *Pecten tissotii* Bernardi, 1858]

1902. *Pecten tissoti* [sic] Bernardi: Yoshihara, *Zool. Mag. Tokyo*, vol. 14, p. 358, pl. 5, fig. 23. [non *Pecton tissotii* Bernardi, 1858]

1912. *Pecten* (*Chlamys*) *bullatus* Dautzenberg and Bavey, Les Lamell. Expéd. Siboga, System., I. Pectinidés, p. 143, pl. 27, figs. 1, 2.

1932. *Chlamys* (*Aequipecten*) *tissotii* (Bernardi): Kuroda, *Venus*, vol. 3, no. 2, p. app. 95. [non *Pecten tissotii* Bernardi, 1858]

1938. *Cryptopecten alli* Dall, Bartsch and Rehder, *Bishop Mus. Bull.*, vol. 153, p. 93, pl. 23, figs. 1–4, 7.

1951. *Cryptopecten tissotii* (Bernardi): Habe, Genera of Japanese Shells, p. 77. [Non *Pecten tissotii* Bernardi, 1858]

1958. *Cryptopecten tissotii* (Bernardi): Habe, *Publ. Seto Mar. Biol. Lab.*, vol. 6, no. 3, p. 265. [non *Pecten tissotii* Bernardi, 1858]

1961. *Cryptopecten tissotii* (Bernardi): Habe, Coloured Illustrations of the Shells of Japan (II), p. 118, pl. 53, fig. 8. [non *Pecten tissotii* Bernardi, 1858]

1964. *Cryptopecten tissotii* (Bernardi): Habe, Shells of the Western Pacific in Color, vol. 2, p. 174, pl. 53, fig. 8. [non *Pecten tissotii* Barnardi, 1858]

1969 *Chlamys* (*Cryptopecten*) *alli* (Dall, Bartsch and Rehder): Hertlein *in* Cox et al., Treatise on Invertebrate Paleontology, pt. N, vol. 1, p. N357, fig. C79–1a, b.

1972. *Cryptopecten* "*tissotii* (Bernardi)": Okutani, *Bull. Tokai Reg. Fish. Res. Lab.*, no. 72, p. 113, fig. 62. [non *Pecton tissotii* Bernardi, 1858]

1975. *Cryptopecten tissotii* (Bernardi): Okutani and Habe, Mollusca II chiefly from Japan, p. 91, 4 figs., p. 256. [non *Pecten tissotii* Bernardi, 1858]

1977. *Cryptopecten tissotii* (Bernardi): Habe, Systematics of Mollusca in Japan, p. 84. [non *Pecten tissotii* Bernardi, 1858]

1979. *Chlamys alli* (Dall, Bartsch and Rehder): Kay, *Bishop Mus. Spec. Publ.* vol. 64, no. 4, p. 524, fig. 168A.

1980. *Aequipecten* sp.: Noda, *Sci. Rep. Inst. Geosci. Univ. Tsukuba*, sec. B, vol. 1, p. 82, pl. 12, fig. 21.

1981. *Cryptopecten alli* Dall, Bartsch and Rehder: Poutiers, *Mem. O. R. S. T. O. M.*, no. 91, p. 326, 332.

1982. *Cryptopecten alli* Dall, Bartsch and Rehder: Hayami, *Venus*, vol. 41, p. 234.

Type.—The holotype of *Pecten* (*Chlamys*) *bullatus* is a conjoined specimen collected by the Siboga expedition at Station 105 off Sulu Archipelago (6°08′N, 121°19′E) in 275 meters. 13.9 mm long, 13.3 mm high, 3.0 mm thick. It is now preserved in the Amsterdam Zoological Museum. The holotype of *Cryptopecten alli* (USNM 173194) is a conjoined specimen collected by the R/V Albatross of the U.S. Bureau of Fisheries at Station 3811 off the south coast of Oahu Island in 238–252 fathoms on coarse sand and rocky bottom. 22.8 mm long, 22.1 mm high, 5.3 mm thick.

Material.—Specimens used for the present description and discussions are indicated in the list of examined samples (p. 122).

Diagnosis.—Medium- or small-sized species of *Cryptopecten* characterized by thin test, weak convexity of both valves, low outline, large apical angle and 18–23 radial ribs which have fine alternate imbricated scales covering both sides of a narrow central ridge and are strongly impressed on internal surface owing to scarcely deposited inner shell layer.

Description.—Shell rarely exceeding 25 mm in length and height, inequivalve, inequilateral, acline in young stage but becoming considerably prosocline in adult stage, left-convex in young but equiconvex or even right-convex in adult; height negatively allometric to length, nearly equal to length in young but invariably less than length in adult; thickness positively allometric to length (and height) in right valve but nearly isometric in left; form ratio T/L scarcely exceeding 0.24 even in adult right valve and 0.18 in adult left valve; length of hinge line moderate, negatively allometric to overall length; test very thin; apical angle of disk about 110 degrees; wings moderate in size, anterior one slightly larger than posterior; pre-umbonal dorsal margin of right valve slightly elevated above hinge axis; posterodorsal margin of disk a little concave and much longer than anterodorsal in adult, while both nearly straight and subequal in length in young; byssal wing moderate in breadth; byssal notch moderate in depth, with four or five exposed denticles of ctenolium and relatively wide fasciole area; anterodorsal and posterodorsal margins of disk scarcely gaped between valves.

Outer surface of disk ornamented with 18–23 radial ribs; each rib consisting of a slender central solid ridge and hollow parts on its sides; solid ridge very sharp, occupying about one-fourth of a rib in breadth; both hollow parts separated into numerous minute chambers and covered with fine imbricated scales numbering about four to six per 1 mm on middle-ventral surface of adult shell; scales on the sides of central ridge very regularly alternate, though oppositely disposed on sides of each interval between ribs; interspace commonly narrower than rib, ornamented with a few fine radial threads only in adult stage, also marked with obliquely inclined scales which occur quite independently from and generally more densely spaced than those on ribs; stepwise growth ring(s) occasionally occur in adult stage; ornamentation on disk essentially the same on both valves; byssal wing commonly possesses five radial ribs which have many spiny scales and become weaker downwards; other wings also have several radial ribs of variable strength.

Coloration variable, but commonly reddish brown and mottled with irregular (oblique or zigzag) pale bands; contrast between parti-colored bands more remarkable in left valve than in right; different pigmentation in two valves also well recognized from inner surface; umbonal area and byssal wing always pale in color.

Inner surface vitreous in luster and strongly plicated in accordance with external sculpture except for umbonal area, due to poor development of inner shell layer; musculature very weakly impressed; resilial insertion relatively small; cardinal crura undeveloped.

Remarks.—Since Yoshihara's (1902) description of several Japanese specimens, many Japanese authors have erroneously called the present species *Pecten tissotii* Bernardi,

1858. This confusion seems to have arisen from Dunker (1882), who assigned a Japanese specimen (not illustrated) to *P. tissoti* [sic] Bernardi with a query. In the original description of *P. tissotii*, Bernardi (1858) did not give any indication about the locality, though *Pecten swiftii* Bernardi, 1858, described in the same article as a collection from the Gulf of Tartary, is a well-known species from north Japan. The original figure of *P. tissotii* (Bernardi, 1858, pl. 1, fig. 2) is at a glace similar to the present species in general outline, but clearly differs from it in the larger number of radial ribs, more sigmoidal frontal margin of left anterior wing, and scarcely impressed radial ribs on the inner surface (suggesting better-developed inner shell layer).

Recently Poutiers (1981) identified a solitary valve sampled by the Musorstom expedition from Lubang of the Philippines (217–230 m in depth) with *Cryptopecten alli*. He said, "The species described and figured by Habe (1964) under the name of '*Cryptopecten tissotii* (Bernardi)' does not correspond to the true *Pecten tissotii* of Bernardi (*J. Conchyl.* 1858, p. 91, pl. 1, fig. 2), the type of which, preserved in the Laboratoire de Malacologie du Museum d'Histoire naturelle de Paris, was examined by the present author" [translated from French]. Moreover, according to Dr. T. R. Waller's personal communication (May 27, 1981), the type specimen of *P. tissotii* in Paris seems to belong to *Aequipecten flabellus* (Gmelin, 1791), which is actually an eastern Atlantic species.

All of the examined specimens from various stations around Japan and the Philippines agree well in every essential character with the Hawaiian specimens of *Cryptopecten alli*, including the holotype. Thus I once applied the specific name of *C. alli* to the species under consideration (Hayami, 1982). Subsequently, however, I recognized that the illustrated specimen (holotype) of *Pecten (Chlamys) bullatus* Dautzenberg and Bavay, 1912, in Amsterdam is nothing but an immature individual of the same species. According to Dr. T. R. Waller's photographs, this specimen shows weak convexity of both valves and about 20 radial ribs, each of which consists of a narrow central ridge and a pair of hollow parts covered with regularly alternate imbricated scales; that is to say, every diagnostic character of *C. alli* is exhibited in this specimen. Consequently, *C. alli* should be regarded as synonymous with *C. bullatus*.

The geographic variation of *C. bullatus* does not seem to be wide; none of the samples from Hawaii, the Philippines or Japan shows any conspicuous local characteristics. In the National Museum of Natural History, however, there is an aberrant small sample (USNM 773983) from Nasca Ridge in the eastern Pacific (25°44′S, 85°25′W, 228 m). All the specimens (three right valves) in this sample are nearly white, and this is certainly not due to abrasion. Because such albinism is rarely met with in samples from the central and northwestern Pacific, this sample may represent an isolated local population.

Fossils of *C. bullatus* seem to be rare. The Late Pleistocene fossil sample *Kk (B)* from the Wan Formation of Kikai Island is composed of somewhat smaller specimens with slightly more numerous radial ribs than Recent specimens. The structure of radial ribs and other characteristics are, however, essentially the same, and subspecific distinction may be unnecessary. A small left valve from the Late Pliocene Shinzato Formation of Okinawa, which was illustrated under the name of *Aequipecten* sp. by Noda (1980), can be regarded as an immature individual of this species.

Distribution.—The present species is widely distributed on the lower sublittoral and upper bathyal sandy substrates in the central and northwestern Pacific. In addition to the type locality and several other stations around the Philippines, it has been known from a number of stations around the Hawaiian and Japanese Islands and a station on Nasca Ridge in the eastern Pacific, as is partly recorded in the list of examined samples. Fossil occurrence of *C. bullatus* is so far restricted to some Pliocene and Pleistocene deposits of the Ryukyu Islands. Late Pliocene (ca. 2.0 Ma) to Recent.

Cryptopecten nux (Reeve)

Diagnosis.—Tumid, smallest species of *Cryptopecten* characterized by relatively strong inflation of both valves, large anterior wing, 18–22 apparently tripartite radial ribs, regularly alternate fine imbricated scales on radial ribs and moderately thick test.

Remarks.—*C. nux* is similar to *C. bullatus* and different from *C. vesiculosus* in the small shell size, large number of radial ribs, slender central ridge of each rib, and regularly alternate disposition of imbricated scales. However, the convexity of both valves is generally much stronger in comparison with the other two species (see Figs. 9, 10). The inner shell layer is poorly developed in *C. bullatus* but attains considerable thickness in the present species as well as in *C. vesiculosus*.

Dr. T. R. Waller suggested to me (oral communication) the possibility that *Pecten nux* Reeve, 1853, together with several subsequently proposed nominal species, may fall into a junior synonym of *Pecten bernardi* Philippi, 1851, because one of the "type-specimens" of *P. bernardi* with no locality data, which is preserved in the British Museum (Natural History) (BM (NH) 1923. 7.13.7), shares essentially the same morphological characters with the illustrated syntype of *P. nux*. As he commented, however, some ambiguity still remains as to the type identity, because Philippi's original description was brief and not accompanied by any illustrations. For the time being I regard *P. nux* as the earliest undoubted name for the species under consideration.

Among several fossil samples of this species from Japan, the Late Pleistocene sample *Sm* (*N*) shows peculiar morphological features deserving a subspecific distinction. Consequently, all the other fossil and Recent samples treated here are provisionally assigned to the nominate subspecies, *Cryptopecten nux nux*.

Geographic variation.—In addition to several fossil and Recent samples from central and south Japan, I have examined numerous well-documented Recent samples of this species from extensive areas of the Indo-Pacific which are preserved in the National Museum of Natural History, Washington D. C. These samples often differ from one another in such characters as shell size, convexity, obliquity, thickness of test, and shape and sculpture of byssal wing. The difference must be partly due to the age heterogeneity but is mostly attributable to geographic variation. The average number and tripartite appearance of radial ribs, however, hardly vary at all. It is my impression that the center of distribution of this species is in the Philippine Islands, though it seems to be also common in some areas of Melanesia, Polynesia, Queensland and the Indian Ocean.

Most of the specimens from the Philippines are characterized by relatively small and right-convex shell and moderate thickness of test. The fossil and Recent samples from various localities in south Japan are not much different from these specimens in shell morphology. On the other hand, the observed specimens from the Marquesas Islands, the type locality of *P. nux*, as well as those from New Britain, show thicker test and more equiconvex shell. The material from Queensland, as exemplified by the specimen USNM 764157 (Plate 9, Fig. 2) and the illustrated syntype of *Chlamys corymbiatus*, possesses thick test and occasionally weak tubercles on the central solid ridge of each radial rib, which are seemingly absent in the specimens from other regions. Generally speaking, the specimens from the Indian Ocean appear to be more weakly inflated than those from the Philippines and many other areas of the Pacific. The holotype of *Chlamys smithi* Sowerby, 1908, from Mauritius Island, seems to exemplify the Indian Ocean form. Moreover, the examined sample (USNM 764165) from the northern part of the Red Sea (eastern coast of Egypt) is unique for the nearly equiconvex and unusually weakly inflated valves (Plate 9, Fig. 5).

The geographic variation of *C. nux* is thus considerably wide. If subspecific division were applied to this species, the trivial names *corymbiatus* and *smithi* (or *guendolenae*) would be available for the populations of Queensland and the main part of the Indian Ocean, respectively. The morphological change in these areas, however, may be gradational, and all the observed samples except a few Philippine ones are too small in sample size to recognize clines or to detect significant morphological difference between local populations.

Cryptopecten nux nux (Reeve)

Pl. 2, Fig. 4; Pl. 3, Figs. 1, 2; Pl. 9, Figs. 2–5; Pl. 12, Figs. 1, 2

1853. *Pecten coruscans* Hinds: Reeve, *Conchologia Iconica*, vol. 8, pl. 32, fig. 143. [non *Pecten coruscans* Hinds, 1845]

1853. *Pecten nux* Reeve, *Conchologia Iconica*, vol. 8, errata.

1888. *Pecten hastingsii* Melvill, *Jour. Conchology*, vol. 5, p. 279, pl. 2, fig. 7.

1888. *Pecten guendolenae* Melvill, *Jour. Conchology*, vol. 5, p. 279, pl. 2, fig. 6.

1888. *Pecten nux* Reeve: Küster and Kobelt *in* Martini and Chemnitz, *System. Conch. Cab.*, vol. 7, pt. 2, p. 163, pl. 45, figs. 5–8.

1908. *Chlamys smithi* Sowerby, *Proc. Malacol. Soc. London*, vol. 8, p. 18, pl. 1, figs. 6, 7.

1909. *Chlamys corymbiatus* Hedley, *Proc. Linn. Soc. New South Wales*, vol. 34, p. 423, pl. 36, figs. 1–4.

1912. *Pecten (Aequipecten) vesiculosus* Dunker: Dautzenberg and Bavay, Les Lamell. Expéd. Siboga, System., 1. Pectinidés, p. 148. [non *Pecten vesiculosus* Dunker, 1877]

1930. *Chlamys (Aequipecten) nux* (Reeve): Cox, *Monogr. Geol. Dept. Hunterian Mus., Glasgow Univ.*, vol. 4, p. 124, pl. 14, fig. 11.

1933. *Pecten (Aequipecten) nux* Reeve: Nomura, *Sci. Rep. Tohoku Imp. Univ.*, ser. 2, vol. 16, no. 1, p. 55.

1934. *Pecten (Aequipecten) nux* Reeve: Nomura and Zinbo, *Sci. Rep. Tohoku Imp. Univ.*, ser. 2, vol. 16, no. 2, p. 117 (pars).

1934. *Pecten (Aequipecten) kikaiensis* Nomura and Zinbo, *Sci. Rep. Tohoku Imp. Univ.*, ser. 2, vol. 16, no. 2, p. 153, pl. 5, fig. 9a, b.

1939. *Corymbichlamys corymbiatus* (Hedley): Iredale, *Brit. Mus. (Nat. Hist.) Sci. Rep.*, vol. 5 (Mollusca pt. 1), p. 368.

1951. *Cryptopecten nux* (Reeve): Habe, Genera of Japanese Shells, p. 77.

1956. *Chlamys (Aequipecten) nux* (Reeve): Eames and Cox, *Proc. Malacol. Soc. London*, vol. 32, p. 43.

1961. *Cryptopecten nux* (Reeve): Habe, Coloured Illustrations of the Shells of Japan (II), p. 118, pl. 53, fig. 9.

1964. *Chlamys (Cryptopecten) nux* (Reeve): Shikama, Selected Shells of the World II, p. 50, fig. 93.

1964. *Cryptopecten nux* (Reeve): Habe, Shells of the Western Pacific in Color, vol. 2, p. 174, pl. 53, fig. 9.

1977. *Cryptopecten nux* (Reeve): Habe, Systematics of Mollusca in Japan, p. 84.

1977. *Cryptopecten kikaiensis* (Nomura and Zinbo): Habe, Systematics of Mollusca in Japan, p. 84.

1977. *Cryptopecten hastingsii* (Melvill): Habe, Systematics of Mollusca in Japan, p. 85.

1982. *Cryptopecten nux* (Reeve): Hayami, *Venus*, vol. 41, p. 235.

Type.—Reeve's illustrated syntype (BM (NH) 1950. 11.14.52), which was erroneously named *"Pecten coruscans"* in the original text but subsequently corrected in the index and errata, is a conjoined specimen from Marquesas Island of Polynesia. It is preserved in the British Museum (Natural History). ca. 15 mm long, 14 mm high, 8 mm thick. The holotype of *Pecten hastingsii* from Japan (ca. 25 mm long, 24 mm high, 14 mm thick) and the holotype of *Pecten guendolenae* from Mauritius (ca. 15 mm long, 14 mm high, 7 mm thick) are conjoined specimens preserved in the Cardiff Museum (Tomlin Collection). The holotype of *Chlamys smithi* from Mauritius (BM (NH) 1908.5.30.63) (ca. 16 mm long, 16 mm high, 7 mm thick) is a conjoined specimen preserved in the British Museum (Natural History). The syntypes of *Chlamys corymbiatus* are said to have been presented to the Australian Museum. The holotype of *Pecten (Aequipecten) kikaiensis* is a right valve (IGPS no. 50357) from the Late Pleistocene Wan Formation at Kamikatetsu of Kikai Island, south Japan, which is preserved in the Institute of Geology and Palaeontology, Tohoku University.

Material.—Specimens used for the following description and discussions are indicated in the list of examined samples (p. 123).

Diagnosis.—Subspecies of *C. nux*, possessing variably but usually strongly inflated valves and narrowly elongated byssal wing with prominent radial ribs and often with spiny scales.

Description.—Shell small, rarely exceeding 20 mm in length and height, highly inequivalve, nearly acline or very slightly prosocline, more or less right-convex except for very early growth stages; height negatively allometric to length, commonly subequal to length in young stages but becoming considerably smaller than length with growth; thickness nearly isometric to length in each valve; form ratio T/L greatly variable even within a sample, ranging 0.23–0.37 in adult right valves and 0.17–0.26 in adult left valves; hinge line moderate or comparatively long, negatively allometric to overall length; test moderate but somewhat variable in thickness; apical angle of disk 100–105 degrees; anterior wing more than twice as long as posterior; byssal wing generally slender, with elevated dorsal margin above hinge axis; anterodorsal and posterodorsal margins of disk nearly straight, subequal in length, scarcely gaped between valves; byssal notch moderate in depth, with four or five exposed denticles of ctenolium and fasciole area of moderate width.

Outer surface of disk ornamented with 18–22 radial ribs, each of which consists of a solid, roof-shaped central ridge and a pair of hollow parts on its sides covered by fine

102

alternate imbricated scales; accordingly, if scales are exfoliated or removed, radial ribs look regularly tripartite; imbricated scales numbering eight to nine per 1 mm on middle-ventral surface of adult shell; interspace of ribs as wide as rib, marked with short erect scales, which are independent from and generally more distantly spaced than imbricated scales on ribs; intercostal radial threads commonly absent but rarely observed in some adult individuals; growth ring(s), if present, very weak; disk sculpture of both valves essentially similar; byssal wing generally possesses four strong radial ribs on which several spiny scales frequently occur; other wings also ornamented with a few radial riblets.

Coloration variable, sometimes monotonously yellow but more commonly reddish brown in ground color with irregular pale bands and spots; left valve generally more darkly pigmented than right, not only externally but also internally; byssal wing always pale in color.

Internal surface somewhat shiny, strongly plicated in accordance with radial ribs, but umbonal to middle part much flattened by deposition of inner shell layer; musculature weakly impressed; resilial insertion moderate in size; cardinal crura developed but not very strong.

Remarks.—This subspecies seems to include almost all the specimens hitherto described under the names of *Pecten nux* and *Chlamys corymbiatus* and also the type materials of *Pecten hastingsii, Pecten guendolenae, Chlamys smithi* and *Pecten (Aequipecten) kikaiensis.*

In spite of an extensive geographic distribution in the Indo-Pacific, the present subspecies seems to be comparatively rare in the seas around the Japanese Islands: every Recent sample consists of one or a few individuals. Only the sample *Kk* (*N*) from the Late Pleistocene of Kikai Island is composed of a large number of individuals, allowing quantitative examination of intrapopulational variation and relative growth (Tables 3, 4, 6–9). At the same locality two other species of *Cryptopecten*, i. e., *C. bullatus* and *C. spinosus*, occur in association, and they are clearly distinguishable from the present subspecies. As suggested from this sample as well as such samples as *Bh* (*N*), USNM 230228, 230314 and 292450 from the Philippines, the intrapopulational variation of shell outline and convexity is considerably wide; the right valve is generally characterized by strong convexity, but the form ratio T/L is quite variable, as indicated by the relatively low value of correlation coefficient between $\log L$ and $\log T$ (or $\log C$).

Besides the sample *Kk* (*N*), *C. nux nux* is represented by several small samples of various geological ages from the Ryukyu Islands (samples *Mk* (*N*), *Ik* (*N*) and *Ob* (*N*)). Though their morphometric comparison is still difficult owing to their small sample size, essential morphological characters including the mode and number of radial ribs appear almost unchanged throughout the geological period.

Neogene fossils of *C. nux* are are considerably well represented in the tropical coastal region of east Africa. Cox (1930) and Eames and Cox (1956) recorded occurrences of this species from the Lower Miocene at several localities in Zanzibar and Pemba Islands in Tanzania, and also from the Pliocene near Mombasa in Kenya. Although I have had no opportunity to observe the African specimens, it is quite likely that the range of this species extends down into the Early Miocene. For the time being I regard these fossils as belonging to *C. nux nux.*

Distribution.—This subspecies is widely distributed in the tropical and subtropical seas of the central-west Pacific and the Indian Ocean. Although dead shells have been dredged from outer sublittoral and bathyal bottoms, the hitherto known records of living shells are almost entirely confined to the inner sublittoral zone. Recent specimens have been known from Polynesia (especially Marquesas and Tahiti), Melanesia, Micronesia, northeast Australia, Indonesia, Singapore, the Philippines, the South China Sea, Formosa, south Japan, Bengal Bay, Andaman, Oman, Seychelles, Mauritius, Madagascar, Mozambique, east coast of South Africa, and the Red Sea. There is no undoubted record of occurrence in the Japan Sea. Fossils of this subspecies occur from the Lower Miocene of east Tanzania (Eames and Cox, 1956), the Pliocene of east Kenya (Cox, 1930), and the Pliocene Byoritsu Formation of Formosa (Nomura, 1933) in addition to the Pliocene and Pleistocene in central and south Japan (see the list of examined samples). Early Miocene to Recent.

Cryptopecten nux sematensis (Oyama)

Pl. 3, Figs. 3, 4

1922. *Pecten tissoti* [sic] Bernardi: Yokoyama, *Jour. Coll. Sci. Imp. Univ. Tokyo*, vol. 44, art. 1, p. 182, pl. 15, figs. 1, 2. [non *Pecten tissotii* Bernardi, 1858]

1954. *Aequipecten* (*Cryptopecten*) *sematensis* Oyama *in* Taki and Oyama, *Palaeont. Soc. Japan, Spec. Papers*, no. 2, pl. 35, figs. 1, 2.

1973. *Aequipecten* (*Cryptopecten*) *sematensis* Oyama: Oyama, *Palaeont. Soc. Japan, Spec. Papers*, no. 17, p. 85, pl. 34, figs. 9a, b, 10a, b.

1976. *Aequipecten vesiculosus* (Dunker): Masuda and Noda, Check List and Bibliography of the Tertiary and Quaternary Mollusca of Japan, p. 8 (only). [non *Pecten vesiculosus* Dunker, 1877]

Type.—The holotype is a nearly complete right valve (UMUT CM21561) which was described by Yokoyama (1922, pl. 15, fig. 1) as "*Pecten tissoti*" and designated by Oyama *in* Taki and Oyama (1954) as the holotype of *Aequipecten* (*Cryptopecten*) *sematensis*. It occurred from "the Upper Musashino at Shito", which is certainly identical with the upper part of the Yabu Formation at Ochishimoshinden of Semata, Toki Town, Chiba Prefecture (35°31.5′N, 140°13.7′E), where the sample *Sm* (*N*) was newly obtained. 15.2 mm long, 17.1 mm high, 6.2 mm thick.

Material.—In addition to the holotype, twelve specimens of the sample *Sm* (*N*) are concerned with the following description and discussions.

Diagnosis.—Subspecies of *C. nux*, characterized by relatively tall outline, very strongly inflated right and left valves, and short and broad byssal wing without any prominent spiny scales.

Description.—Shell small, rarely exceeding 18 mm in length and height, nearly acline throughout growth, right-convex except for very young stages; height isometric or slightly negatively allometric to length, commonly exceeding length; thickness decidedly positively allometric to length (and height) in each valve; form ratio T/L as large as 0.36–0.41 in adult right valve and 0.26–0.29 in adult left valve; apical angle of disk about 100–105 degrees; test relatively thick; byssal wing broad, not much elongated, marked with several (commonly five) weak radial ribs without development of any prominent spiny scales; disk sculpture and other morphological characters essentially similar to those of *C. nux nux*.

Remarks.—When Oyama *in* Taki and Oyama (1954) made a taxonomic revision on Yokoyama's illustrated molluscs from the Quaternary of Kanto region, a new name, *Aequipecten* (*Cryptopecten*) *sematensis*, was introduced for two tumid specimens of a pectinid which had been referred to *Pecten tissoti* [error of *tissotii*] by Yokoyama (1922).

Two different views are possible as to the validity of Oyama's proposal of this taxonomic name. One interpretation, as pointed out by Hayami (1973) and Masuda and Noda (1976), is that *A.* (*C.*) *sematensis* Oyama, 1954, is a *nomen nudum*, because the proposal was not accompanied by any indication of diagnostic characters. Under this interpretation the present description would constitute the first valid proposal. The other view, as personally communicated by Dr. K. Oyama (October 16, 1982), is that the taxonomic name has been available since 1954, because the citation [synonym list first appeared in Oyama (1973)] of Yokoyama's description of "*Pecten tissoti* [sic]" constitutes and valid indication as required by ICZN articles 13 (a) (ii) and 16 (a) (i). I am now in favor of the latter interpretation, applying Oyama's trivial name, *sematensis*, for this taxon.

My examination of one of Yokoyama's specimens (UMUT CM21561) (the other specimen CM21562, which was erroneously designated as the holotype of *A.* (*C.*) *sematensis* by Oyama (1973), is missing from the University Museum, the University of Tokyo), and also of a newly collected sample *Sm* (*N*), showed that the structure of radial ribs and many other characteristics are quite similar to those of *C. nux* from other fossil localities and the present seas. The convexity of the two valves is, however, unusually strong, and the thinkness (T) is far more positively allometric to length (L) and height (H). The difference in the shell convexity and growth ratio (a) between this subspecies and *C. nux nux* is clearly recognizable from the profiles of the two valves (Pl. 3, Fig. 1c versus Fig. 3c; Fig. 2c versus Fig. 4c) as well as the discrepant slopes of reduced major axes in the double logarithmic diagrams of relative growth (see Figs. 15 and 16). Another criterion for the subspecific distinction seems to exist in the shape and sculpture of the byssal wing. In the present subspecies the byssal wing is not so narrowly elongated nor so strongly ornamented as in the nominate subspecies.

Distribution.—This subspecies has been known almost exclusively from the Yabu Formation at Ochishimoshinden of Semata, Chiba Prefecture. I have found only one left valve of this subspecies in the old collection of the Tohoku University (IGPS no. 13369), a specimen which was collected by Prof. Yabe from "Narita Beds at Takakura, Shito Village" and registered as *Pecten* (*Chlamys*) *vesiculosus*. Late Pleistocene (ca. 0.29 Ma).

Cryptopecten vesiculosus (Dunker)

Diagnosis.—Medium- or large-sized species of *Cryptopecten* characterized by relatively thick test, moderately strong convexity of both valves, tall outline, small apical angle and 13–18 radial ribs, each of which consists of a flat-topped central solid ridge and a pair of hollow parts covered by oppositely disposed imbricated scales.

Remarks.—*C. vesiculosus* is morphologically distinct from *C. bullatus* and *C. nux* in the larger shell size, much fewer radial ribs and oppositely disposed imbricated scales on the sides of each rib. The shell convexity is generally stronger than that of *C. bullatus* and weaker than that of *C. nux*. In accordance with the present proposal of a new subspecies for two Pliocene fossil samples from the Kakegawa Group of central Japan, all the Recent and other fossil samples of this species are assigned to its nominate sub-species, *Cryptopecten vesiculosus vesiculosus*.

Cryptopecten vesiculosus vesiculosus (Dunker)

Pl. 3, Fig. 5; Pl. 4, Figs. 1–3; Pl. 5, Figs. 1–3; Pl. 6, Figs. 1–6; Pl. 7, Figs. 1–10; Pl. 10, Figs. 1, 2; Pl. 11, Figs. 1, 2; Pl. 12, Fig. 12; Pl. 13, Figs. 1, 2

1877. *Pecten vesiculosus* Dunker, *Malakol. Bl.*, vol. 24, p. 72.

?1877. *Pecten trifidus* Dunker, *Malakol. Bl.*, vol. 24, p. 72.

1882. *Pecten vesiculosus* Dunker: Dunker, Index Molluscorum maris Japonici, p. 241, pl. 11, fig. 1 [Phenotype *Q*].

?1882. *Pecten jickelii* Dunker, Index Molluscorum maris Japonici, p. 241. [new name for *Pecten trifidus* Dunker, 1877].

1888. *Pecten vesiculosus* Dunker: Küster and Kobelt *in* Martini and Chemnitz, *System. Conch. Cab.*, vol. 7, p. 138, pl. 38, fig. 4 [Phenotype *Q*].

?1888. *Pecten hysginodes* Melvill, *Jour. Conch.*, vol. 5, p. 280, pl. 2, fig. 8 [? Phenotype *R*].

1902. *Pecten vesiculosus* Dunker: Yoshihara, *Zool. Mag. Tokyo*, vol. 14, p. 212, pl. 4, fig. 16 [Phenotype *Q*].

1911. *Pecten vesiculosus* Dunker: Yokoyama, *Jour. Geol. Soc. Tokyo*, vol. 18, p. 1, pl. 1, figs. 8–10 [Phenotype *Q*].

1920. *Pecten vesiculosus* Dunker: Yokoyama, *Jour. Coll. Sci. Imp. Univ. Tokyo*, vol. 39, art. 6, p. 154, pl. 13, figs. 11–13 [Phenotype *Q*].

1922. *Pecten vesiculosus* Dunker: Yokoyama, *Jour. Coll. Sci. Imp. Univ. Tokyo*, vol. 44, art. 1, p. 182.

1925. *Pecten vesiculosus* Dunker: Yokoyama, *Jour. Coll. Sci. Imp. Univ. Tokyo*, vol. 45, art. 5, p. 27.

1928. *Pecten vesiculosus* Dunker: Yokoyama, *Jour. Fac. Sci. Imp. Univ. Tokyo*, sec. 2, vol. 2, pt. 7, p. 349.

1929. *Chlamys (Aequipecten) vesiculosus* [sic] (Dunker): Fujita, *Venus*, vol. 1, no. 1, p. 60.

1932. *Chlamys (Aequipecten) vesiculosus* [sic] (Dunker): Kuroda, *Venus*, vol. 3, no. 2, p. 94 (app.), fig. 106 [Phenotype *Q*], fig. 107 [Phenotype *R*].

1934. *Chlamys (Aequipecten) vesiculosa* (Dunker): Hirase, Collection of Japanese Shells, p. 8, 111, pl. 12, fig. 6 [Phenotype *Q*].

1951. *Cryptopecten vesiculosus* (Dunker): Habe, Genera of Japanese Shells, p. 77, figs. 155–158.

1951. *Cryptopecten vesiculosus* (Dunker): Taki *in* Hirase, Illustrated Handbook of Shells, pl. 12, fig. 6 [Phenotype *Q*].

1954. *Aequipecten (Cryptopecten) vesiculosus* (Dunker): Taki and Oyama, *Palaeont. Soc. Japan, Spec. Papers*, no. 2, pl. 14, figs. 11–13 [Phenotype *Q*] [reproduction of Yokoyama's (1920) figures].

1954. *Cryptopecten vesiculosus* (Dunker): Kira, Coloured Illustrations of the Shells of Japan, p. 123, pl. 49, fig. 1 [Phenotype *Q*].

1958. *Cryptopecten vesiculosus* (Dunker): Habe, *Publ. Seto Mar. Biol. Lab.*, vol. 6, no. 3, p. 265.

1958. *Chlamys vesiculosa* (Dunker): Shirai, *Venus*, vol. 20, no. 1, p. 88.

1960. *Cryptopecten vesiculosus* (Dunker): Shuto, *Mem. Fac. Sci. Kyushu Univ.*, ser. D, vol. 9, no. 3, p. 123, pl. 13, fig. 3 [Phenotype *Q*].

1962. *Aequipecten vesiculosus* (Dunker): Masuda, *Sci. Rep. Tohoku Univ.*, ser. 2, vol. 33, no. 2, p. 191, pl. 24, figs. 7, 8, pl. 25, fig. 7 [Phenotype *Q*].

1967. *Cryptopecten vesiculosus* (Dunker): Habe and Kosuge, Shells, p. 134, pl. 50, fig. 12 [Phenotype *R*], fig. 13 [Phenotype *Q*].

1968. *Aequipecten vesiculosus* (Dunker): Ohara, Geol. Atlas Chiba Pref., no. 5. Bivalvia, pl. 3, fig. 6a, b [Phenotype *Q*].

1969. *Aequipecten (Cryptopecten) vesiculosus* (Dunker): Shikama and Masujima, *Sci. Rep. Yokohama Nat. Univ.*, ser. 2, no. 15, pl. 7, fig. 14 [Phenotype *Q*].

1971. *Cryptopecten vesiculosus* (Dunker): Kuroda, Habe and Oyama *in* Biological Laboratory, Imperial Household, Seashells of Sagami Bay, p. 574/366, pl. 79, figs. 1–4 [Phenotype *Q*], figs. 5–7 [Phenotype *R*].

1972. *Cryptopecten vesiculosus* (Dunker): Hayami, *Jour. Geol. Soc. Japan*, vol. 78, no. 9, p. 495, 499, text-fig. 1 (left) [Phenotype *Q*], text-fig. 1 (right) [Pehnotype *R*].

1973. *Cryptopecten vesiculosus* (Dunker): Hayami, *Jour. Paleontology*, vol. 47, no. 3, p. 401, pl. 1, figs. 1–4, 9, 10, pl. 2, figs. 1, 2, 7, 8 [Phenotype *R*], pl. 1, figs. 5–8, 11, 12, pl. 2, figs. 3–6, 9–12 [Phenotype *Q*].

1973. *Aequipecten vesiculosus* (Dunker): Hayasaka, *Sci. Rep. Tohoku Univ.* ser. 2, spec. vol., no. 6, p. 102, pl. 6, fig. 3 [Phenotype *Q*].

1973. *Aequipecten (Cryptopecten) vesiculosus* (Dunker): Oyama, *Palaeont. Soc. Japan, Spec. Papers*, no. 17, p. 85, pl. 34, figs. 1–3 [Phenotype *Q*] [reproduction of Yokoyama's (1920) figures].

1975. *Cryptopecten vesiculosus* (Dunker): Okutani and Habe, Mollusca II chiefly from Japan, p. 90, 4 figs. [Phenotype *R*], p. 255.

1975. *Cryptopecten vesiculosus* (Dunker): Hayami and Ozawa, *Lethaia*, vol. 8, p. 10, fig. 9 (left) [Phenotype *R*], fig. 9 (right) [Phenotype *Q*].

1977. *Cryptopecten vesiculosus* (Dunker): Habe, Systematics of Mollusca in Japan, p. 84, pl. 16, figs. 3–5.

1979. *Cryptopecten vesiculosus* (Dunker): Nemoto and Ohara, Taira Chigaku Dokokai Kaiho, spec. vol., p. 56, pl. 3, fig. 3 [Phenotype *Q*].

1979. *Aequipecten (Cryptopecten) vesiculosus* (Dunker): Mori and Osada, *Mater. Rep. Hiratsuka City Museum*, no. 19, p. 38, pl. 8, fig. 10 [Phenotype *Q*].

1979. *Aequipecten (Cryptopecten) sematensis* Oyama: Mori and Osada, *Mater. Rep. Hiratsuka City Museum*, no. 19, p. 38, pl. 8, fig. 11 [Phenotype *R*] [non *Aequipecten (Cryptopecten) sematensis* Oyama, 1954].

1982. *Cryptopecten vesiculosus* (Dunker): Hayami, *Venus*, vol. 41, p. 235.

Type.—The original description of *Pecten vesiculosus* by Dunker (1877) was not accompanied by any illustrations, but a specimen subsequently figured by the same author (Dunker, 1882, pl. 11, fig. 1) is presumably one of the syntypes. The locality mentioned was Japan, but details are unknown.

Material.—Numerous specimens, as indicated in the list of examined samples (p. 127), were used for the present description and discussions.

Diagnosis.—Medium-sized subspecies of *Cryptopecten vesiculosus* characterized by relatively small number of radial ribs, thick test with well-developed inner shell layer and broad central solid ridge of radial rib. Middle Pleistocene and later populations remarkably dimorphic in surface sculpture. This subspecies occurs predominantly on coarse-grained substrates and in sandy sediments.

Description.—Shell rarely exceeding 34 mm (38 mm in fossils) in length and height, highly inequivalve, inequilateral, nearly acline throughout growth, left-convex in young stages but becoming right-convex in adult; height always negatively allometric to length, commonly surpassing length in young but subequal to or rather smaller than length in adult; thickness positively allometric to length (and height) in right valve but commonly negatively allometric in left; form ratio T/L as large as 0.25–0.30 in adult right valve and 0.20–0.24 in adult left valve; length of hinge line moderate, negatively allometric to overall length; test comparatively thick; apical angle of disk 90–95 degrees in Recent specimens but often exceeding 95 degrees in Pliocene fossils; anterior wing about twice as long as posterior; byssal wing comparatively broad, with elevated dorsal margin above hinge axis; posterodorsal margin of disk slightly concave and slightly longer than antero-dorsal in adult, but subequal in length in young; byssal notch moderately deep, with three or four exposed denticles of ctenolium and relatively narrow fasciole area; antero-dorsal and posterodorsal margins of disk scarcely gaped between two valves.

Outer surface of disk ornamented with 13–18 broad, regular, radial ribs; structure of radial sculpture quite different in two phenotypes, as separately described; very conspic-uous stepwise growth ring(s) frequently occur in adult individuals; byssal wing com-monly ornamented with five (sometimes four or six) strong radial ribs, on which a number of finny scales occur; other wings also ornamented with several radial riblets, upper one or two of which are commonly stronger than the rest.

Coloration variable, but commonly reddish brown in ground color, often variegated with pale bands and spots; contrast between darkly pigmented part and pale part more remarkable in left valve than in right; byssal wing and early dissoconch of both valves almost invariably pale in color.

Inner surface porcellaneous in luster except for vitreous muscle portion owing to well-developed inner foliate layer, almost white in right valve but more or less darkly pig-mented in left valve; radial ribs weakly impressed on central-ventral area; adductor muscle scar clearly impressed, moderate in size, consisting of striate (quick) muscle portion and smooth (slow) muscle portion, which, however, are obscurely discriminated in left valve; position of striate muscle scar somewhat different between two valves owing to oblique direction of quick muscle; resilial insertion moderate in size, bordered by a pair of strong cardial crura.

Sculpture difference in two phenotypes.—The radial ribs on the disk are highly elevated and subquadrate in transverse section in Phenotype Q, whereas they are rounded and very low in Phenotype R. The central solid ridge of each radial rib is flat-topped and comparatively wide in both phenotypes. Its surface is always smooth in Phenotype R, but may or may not have fine erect scales in Phenotype Q. In both phenotypes the central ridge is broader in the right valve than in the left. In Phenotype Q, however, the sculpture is essentially the same in both valves, the lateral slopes of the solid ridge being steep and forming a pair of hollow parts which are covered by imbricated scales numbering about three per 1 mm on the middle-ventral surface of adult shell. In the same phenotype the interspace between ribs possesses fine erect scales which are never imbricated and more distantly spaced than the scales on the hollow parts. On the other hand, in Phenotype R the radial sculpture is considerably different in the two valves: in the right valve the inter-space is quite narrow in the young stage and not clearly discriminated from the slopes of ribs in the adult, while in the left valve it is wide and completely covered by imbricated scales numbering about eight per 1 mm on the middle-ventral surface of adult shell, which are quite analogous with those on the hollow parts, not only in shape but also in periodicity. In other words, hollow parts are restricted to the lateral sides of the central ridge in Phenotype Q, whereas imbricated scales also cover the entire rib interspace of the left valve in Phenotype *R*. In the right valve, the hollow parts are persistent in Phenotype *Q* but have completely vanished on the ventral surface in Phenotype *R*. In transverse section, the hollow space is lenticular in Phenotype *Q* but subcircular in Phenotype *R*. In both phenotypes the hollow space is always wider in the left valve than in the right. Radial ribs on the wings are invariably solid without any hollow space, and their structure is essentially the same in the two phenotypes.

Remarks.—Cryptopecten vesiculosus is a common pectinid in Japan, but its morphology and variation have not been studied in detail. The remarkable discontinuous variation of disk sculpture, however, seems to have been noticed by some previous authors. For example, Kuroda (1932) illustrated two left valves belonging to different phenotypes in juxtaposition. Furthermore, Habe and Kosuge (1967) and Kuroda, Habe and Oyama *in* Biological Laboratory, Imperial Household (1971) mentioned, though briefly, the presence of dimorphic phenomenon in their materials. Fortunately, however, except for Kira (1950) and Mori and Osada (1979), no one seems to have regarded the two phenotypes as separable into different taxa. As already documented in this paper, the two phenotypes coexist in every large sample after the Middle Pleistocene and are surely attributable to intrapopulational discontinuous variation.

Some problems about the synonymy of early proposed specific names remain unsolved. *Pecten trifidus* Dunker, 1877, from Japan, which was renamed *Pecten jickelii* by Dunker (1882) without any adequate reason, may be conspecific with *P. vesiculosus*, as was suggested by Habe (1977). The original descriptions of *P. trifidus* and *P. jickelii* were not accompanied by any illustrations, but the number of radial ribs, which was said to be 18 in *P. trifidus*, may be somewhat larger than the averages in the present samples of *C. vesiculosus*. Their taxonomic position, therefore, is difficult to determine without examining the type material. On the basis of my observation of Dr. T. R. Waller's unpublished photographs, I conclude that the type conjoined specimen of *Pecten hysginodes* Melvill, 1888, from unknown locality, which is preserved in the Cardiff Museum, may belong to the Phenotype *R* of *C. vesiculosus*.

The detailed morphology and phenotypic frequency are, as described in the preceding chapters, considerably variable in time and space, but all the samples here treated certainly belong to a single evolutionary species. Masuda (1962) mentioned the existence of some morphological difference between the Pliocene and Pleistocene-Recent specimens of this species. He pointed out that some intercostal threads are developed in the Recent and Pleistocene specimens near the ventral margin, but that they are usually absent in Tertiary ones. Furthermore, the Pliocene specimens were said to have broader ribs and narrower interspaces than Pleistocene-Recent ones. My observation of the present samples of various ages revealed these characters to be more closely related to phenotypic and subphenotypic differences, ontogenetical change and state of preservation than to chronological change. The intercostal sculpture actually differs not only between the two phenotypes but also between the two subphenotypes (p. 58) within Phenotype *Q*; intercostal threads commonly appear in adult individuals of *rough* subphenotype but are rarely met with in those of *smooth* subphenotype almost regardless of geological horizons. Many well-preserved large specimens from the Pliocene, for example, those in samples *Sh* 1 and *Ik*, show as clear intercostal threads as those of Pleistocene-Recent samples, while smaller specimens always lack such threads. The relative breadth of radial ribs varies considerably among individuals within one sample and increases remarkably with growth. A striking feature is that radial ribs appear very narrow if imbricated scales on their sides are exfoliated. Though biometrical comparison of this character among samples has not been carried out owing to the different growth stages and difficulty of subphenotypic discrimination in many unfavorably preserved samples, chronological change could not be detected with respect to the relative breadth of ribs and interspaces.

Distribution.—This subspecies lives on the outer sublittoral sandy substrates around the central and southwestern parts of the Japanese Islands under the influence of Kuroshio warm current (35°N and south) and its branch, Tsushima Current (41°N and south) (see Figs. 1 and 3). Fossils of this subspecies are also almost entirely restricted to the same area as the distribution in the present time, though the northern limit seems to have fluctuated with time (see Fig. 4). Middle Pliocene (ca. 3.5 Ma) to Recent.

Cryptopecten vesiculosus makiyamai subsp. nov.

Pl. 7, Figs. 11, 12

Type.—The holotype is a left valve (UMUT CM 16033a) from the Hosoya Silt of the Kakegawa Group at the west of Ugari, northeast of Yamanashi, Fukuroi City, Shizuoka Prefecture, central Honshu (34°47.5′N, 137°56.3′E). 44.5 mm long, 43.1 mm high, 9.9 mm thick.

Material (Paratypes).—In addition to the holotype 25 specimens of the samples *Kg 1* from the type locality and *Kg 2* from a nearby locality are concerned with the following description and discussions. See also the list of examined samples (p. 127).

Diagnosis.—Unusually large-sized subspecies of *Cryptopecten vesiculosus* characterized by rapid growth rate, relatively large number of radial ribs, thin test and narrow central solid ridge of radial rib. This subspecies characteristically occurs in silty sediments.

Discription.—Shell often exceeding 40 mm in length and height; test comparatively thin; apical angle approximately 95 degrees; 15–18 radial ribs, with an average of 16, each of which consists of a comparatively narrow central ridge and a pair of relatively wide hollow parts; rib interspace marked with a number of fine radial threads in adult shell, showing lattice ornament with fine erect scales; other characteristics essentially similar to those of Phenotype *Q* of *C. vesiculosus vesiculosus*.

Remarks.—Since Makiyama's (1931) stratigraphic work, it has been known that *C. vesiculosus* occurs frequently in the Hosoya Silt of the Kakegawa Group in Ugari area of Fukuroi City. No paleontological description, however, has been made on the material of this member.

Owing to the small sample size and appreciable secondary deformation, detailed morphometric comparison with other samples is still difficult. The specimens from the Hosoya Silt (samples *Kg 1* and *Kg 2*) are, however, morphologically and paleoecologically very unique in the following points. Firstly, the maximum shell size and growth rate are unusually large as shown in Table 13 and Fig. 7. In spite of the large size, the test is comparatively thin. Secondly, the central solid ridge of each rib is much narrower, and the hollow parts, though the covering imbricated scales have been lost in every specimen, are somewhat wider than in other samples of *C. vesiculosus*.Thirdly, the average number of radial ribs on the disk is slightly but significantly larger than that in any other sample of this species, as indicated in Table 11. Fourthly, as discussed already (p. 80), the habitat and associated molluscs seem to be considerably different from those of Recent and other fossil populations of *C. vesiculosus*. This subspecies is found exclusively in silty beds. Because the age of the Hosoya Silt is certainly within the range of *C. vesiculosus*, I provesionally regard this phenon as constituting a geographical (or ecological) subspecies.

Distribution.—This subspecies is known only from the Hosoya Silt Member of the Kakegawa Group in central Honshu, Japan. Late Pliocene (1.9–2.4 Ma).

Cryptopecten phrygium (Dall)

Pl. 9, Figs. 6–9

1886. *Pecten phrygium* Dall, *Bull. Mus. Comp. Zool., Harvard Coll.*, vol. 12, no. 6, p. 217, pl. 40, fig. 1.
1954. *Aequipecten* (*Aequipecten*) *phrygium* Dall: Abbott, American Seashells, p. 367.
1974. *Aequipecten phrygium* (Dall): Abbott, American Seashells, 2nd ed., p. 445.
1982. *Cryptopecten phrygium* (Dall): Woodring, *U. S. Geol. Surv. Prof. Paper*, 306-F, p. 595.

Type.—The illustrated syntype is a living conjoined specimen dredged at Station 32 of the U.S. Coast Survey Steamer Blake in 95 fathoms, north of Yucatan Banks (23°32′N, 88°05′W) in the Mexican Gulf. 36.5 mm long, 36.5 mm high. Several dead shells, which also constitute the syntypes, were collected at three other stations in the same gulf. These specimens are preserved in the Museum of Comparative Zoology, Harvard University.

Material.—Several small samples in the National Museum of Natural History, Washington D. C., and the American Museum of Natural History, New York, are concerned with the following description and discussions, as indicated in the list of examined samples (p. 137).

Diagnosis.—Large-sized species of *Cryptopecten* characterized by weak convexity of both valves, relatively short hinge line, and 17–19 radial ribs with sharply angulated central solid ridge and regularly alternate imbricated scales.

Description.—Shell sometimes exceeding 45 mm in length and 43 mm in height, inequivalve, inequilateral, nearly acline in young stages but considerably prosocline in adult, relatively tall in young but rather long in adult, equiconvex or slightly right-convex; test moderate in thickness; hinge line decidedly short in adult stages, though rather long in young; posterodorsal margin of disk broadly concave, much longer than anterodorsal; apical angle of disk about 100 degrees; wings relatively small in adult stages; byssal notch moderately deep, provided with about five exposed denticles of ctenolium and narrow fasciole area.

Outer surface of disk ornamented with 17–19 highly elevated radial ribs, each of which is reinforced by an angular solid ridge; hollow parts on both sides of ridges covered with alternately disposed imbricated scales, which number about five per 1 mm on middle-ventral surface of adult shell; rib interspace moderate in width, marked with fine scales and in adult stage, with a few fine intercostal threads as well; both wings ornamented with several weak radial riblets; growth rings, if present, rather inconspicuous.

Coloration considerably variable, as in other species of this genus, and commonly mottled with reddish ground and irregular pale spots.

Inner surface porcellaneous in luster owing to well-developed inner foliated layer, almost white in right valve but more or less darkly pigmented in left valve; radial ribs impressed on central-ventral area; muscle impression similar to that of *C. vesiculosus*; cardinal crura comparatively weak.

Remarks.—*Cryptopecten phrygium* is a solitary representative of this genus in the Atlantic. It may be an uncommon species, judging from the scarceness of subsequently described specimens. In the National Museum of Natural History and the American Museum of Natural History, there are many small samples which agree well with the original description, but morphometrical treatment of the intrapopulational and geographic variations and allometric growth is difficult owing to the small number of individuals. Nevertheless, a highly negatively allometric growth of H to L as well as of D to L is strongly suggested. In short, every essential character indicates that this species is closely related to Pacific species of *Cryptopecten*.

It resembles *C. bullatus* in the weak convexity of both valves and the mode of radial sculpture characterized by the sharp central ridge and regularly alternate imbricated scales on the hollow parts. However, the maximum shell size is almost twice as large as that of *C. bullatus*. If specimens of similar size from the two species are compared, it is easy to see that *C. phrygium* has much taller shell than *C. bullatus*. Moreover, the average number of radial ribs is decidedly fewer in the present species. In the developed inner layer and clearly impressed muscle scars it may be rather similar to *C. vesiculosus*, but clearly differs from the Japanese species in the more numerous radial ribs, narrower central ridge, weaker shell convexity and alternate disposition of imbricated scales.

Distribution.—Living specimens of this species are known from a depth of about 200 meters. In addition to many stations in the Gulf of Mexico, this species has been known from the Atlantic coast (from Cape Cod to Florida), the Lesser Antilles and a station off Georgetown in Guiana. Recent.

Cryptopecten spinosus sp. nov.

Pl. 8, Figs. 1–5; Pl. 12, Figs. 1, 2

1934. *Pecten (Aequipecten) vesiculosus* Dunker: Nomura and Zinbo, *Sci. Rep. Tohoku Imp. Univ.*, ser. 2, vol. 16, no. 2, p. 117. [non *Pecten vesiculosus* Dunker, 1877]

Type.—The holotype is a nearly complete right valve (UMUT CM16170c) from the coral sand of the Late Pleistocene Wan Formation of the Ryukyu Group at about 500 m north of Kamikatetsu, Kikai Island (Kikai Town), Kagoshima Prefecture, south Japan (28°17.0′N, 129°56.8′E). 25.7 mm long, 25.1 mm high, 6.1 mm thick.

Material (Paratypes).—In addition to the holotype, 189 specimens of the sample *Kk* (*S*) from the type locality are concerned with the following description and discussions.

Diagnosis.—Medium-sized species of *Cryptopecten*, characterized by relatively weak convexity of left valve, 12–15 radial ribs, a narrow and sharp central ridge which is spinously projected beyong ventral margin, somewhat coarse and oppositely disposed imbricated scales covering tubular hollow parts, and conspicuous scales on byssal wing.

Description.—Shell rarely exceeding 26 mm in length and height, highly inequivalve, subequilateral, nearly acline throughout growth, right-convex except for very early stages of dissoconch; height commonly exceeding length in young stages, but negatively allometric and becoming subequal to length in adult stages; thickness positively allo-

112

metric to length in right valve but isometric or rather negatively allometric in left valve; form ratio T/L as small as 0.23–0.27 in adult right valve and 0.17–0.19 in adult left valve; hinge line comparatively short, negatively allometric to overall length; test solid, moderate in thickness; wings highly unequal in size; pre-umbonal dorsal margin of right valve elevated above hinge axis; posterodorsal margin of disk slightly concave, a little longer than anterodorsal; apical angle about 90–95 degrees; byssal notch considerably deep, provided with three or four exposed denticles of ctenolium and relatively narrow fasciole area.

Outer surface of disk ornamented with 12–15 highly elevated and strong radial ribs; each rib reinforced by a narrow, flat-topped and scaly central solid ridge, the sides of which are steep and remarkably excavated; imbricated scales oppositely disposed, coarse (about four per 1 mm), strongly convex and granular in appearance; consequently, hollow space covered with imbricated scales comparatively wide and tubular; terminal of each central ridge spinously projected beyond ventral margin and observable also from inside of shell; rib interspace moderate in width, marked with erect scales which are quite independent from and more distantly spaced than imbricated scales on hollow parts; fine intercostal threads occurring only in adult stages; byssal wing ornamented with four (occasionally five) radial riblets on which several finny scales are developed; other wings marked with weaker riblets; stepwise growth rings sometimes observed on adult shell.

Color pattern recognizable, though obscurely, on external and internal surface especially in left valve, characterized by irregular bands and spots on darkly pigmented ground. Internal surface covered with thin inner shell layer with clearly impressed radial sculpture except for umbonal area; cardinal crura relatively weak; muscle scars not much different from those of *C. vesiculosus vesiculosus* in size and shape.

Remarks.—The present species, as interpreted by Nomura and Zinbo (1934), resembles *C. vesiculosus*, but is clearly distinguishable from that species by the weaker convexity of both valves (see Figs. 15 and 16), significantly fewer radial ribs (Table 10) and central ridge projected beyond ventral margin. One of the most distinctive characters is the shape of the imbricated scales on the hollow parts of the ribs. In Phenotype Q of *C. vesiculosus* the slopes of the central solid ridge are never so deeply excavated, and the imbricated scales are not so strongly convex. Consequently, in the transverse section and also in ventral view, the hollow space is lenticular in Phenotype Q of *C. vesiculosus* but nearly circular in the present species. In short, the sculpture of *C. spinosus* is more prominent and decorative than that of *C. vesiculosus* and other species of *Cryptopecten*.

The present new species is known only from raised reef sediments of the Late Pleistocene age, and appears to have soon become extinct. Its reproductive isolation from *C. vesiculosus* before this stage is suggested by the absence of dimorphism in the sample *Kk* (*S*). I presume that *C. spinosus* represents an incipient species which was derived from the main stock of *C. vesiculosus*.

Distribution.—The present species is known only from the type locality. According to Omura (1983), this fossil bed represents grainstone facies deposited at a depth of 40–100 meters, and its $^{230}\mathrm{Th}/^{234}\mathrm{U}$ age is 82,000±2,000 years B. P. as determined from three species of ahermatypic corals. Late Pleistocene.

Cryptopecten yanagawaensis (Nomura and Zinbo)

Pl. 8, Figs. 6–9

1936. *Pecten (Aequipecten?) yanagawaensis* Nomura and Zinbo, *Saito Ho-on Kai Mus. Res. Bull.*, no. 10, p. 337, pl. 20, fig. 2a, b.

1940. *Pecten (Aequipecten) yanagawaensis* Nomura and Zinbo: Nomura, *Sci. Rep. Tohoku Imp. Univ.*, ser. 2, vol. 21, no. 1, p. 19, pl. 1, figs. 10–13.

1958. *Cryptopecten yanagawaensis* (Nomura and Zinbo): Masuda, *Trans. Proc. Palaeont. Soc. Japan*, n. s., no. 30, p. 189, pl. 27b, figs. 1–8.

1962. *Aequipecten yanagawaensis* (Nomura and Zinbo): Masuda, *Sci. Rep. Tohoku Univ.*, ser. 2, vol. 33, no. 2, p. 192, pl. 26, fig. 8.

1965. *Aequipecten yanagawaensis* (Nomura and Zinbo): Masuda and Takegawa, *Saito Ho-on Kai Mus. Res. Bull.*, no. 40, pl. 1, figs. 12, 13.

?1973. *Aequipecten (Cryptopecten) yanagawaensis* (Nomura and Zinbo): Shikama, *Sci. Rep. Tohoku Univ.*, ser. 2, spec. vol., no. 6, p. 190, 194.

1973. *Aequipecten yanagawaensis* (Nomura and Zinbo): Masuda, Atlas of Japanese Fossils, no. 33, pl. N-54, figs. 14–16.

1974. *Cryptopecten yanagawaensis* (Nomura and Zinbo): Itoigawa, Shibata and Nishimoto, *Bull. Mizunami Fossil Mus.*, no. 1, p. 67, pl. 11, figs. 6–9.

1976. *Aequipecten yanagawaensis* (Nomura and Zinbo): Ogasawara, *Sci. Rep. Tohoku Univ.*, ser. 2, vol. 46, no. 2, p. 44, pl. 3, figs. 3, 6.

1981. *Aequipecten yanagawaensis* (Nomura and Zinbo): Itoigawa, Shibata, Nishimoto and Okumura, *Monogr. Mizunami Fossil Mus.*, no. 3A, pl. 7, figs. 2, 3.

1982. *Aequipecten yanagawaensis* (Nomura and Zinbo): Itoigawa, Shibata, Nishimoto and Okumura, *Monogr. Mizunami Fossil Mus.*, no. 3B, p. 46.

1982. *Cryptopecten yanagawaensis* (Nomura and Zinbo): Hayami, *Venus*, vol. 41, no. 3, p. 235.

Type.—The holotype in Saito Ho-on Kai Museum, Sendai (SM 8353) was collected from the early Middle Miocene Yanagawa Formation at a river cliff of the Hirose-gawa, the southwest end of Yanagawa Park, Yanagawa Town, Fukushima Prefecture, north Japan (37°51.1′N, 140°36.1′E). 21.5 mm long, 20 mm high, 5 mm thick.

Material.—The following description is based on the sample *Mn (Y)* from the early Middle Miocene Moniwa Formation of the Natori Group at the southwest of Jûnishin of Natori City, Miyagi Prefecture, north Japan. The type and other material of this species in the Saito Ho-on Kai Museum and Tohoku University were also examined.

Diagnosis.—Medium-sized species of *Cryptopecten* characterized by relatively weak convexity of both valves, low outline, large apical angle, 20–25 radial ribs with broad and flat-topped central ridge, oppositely disposed imbricated scales on ribs, strong cardinal crura and scarcely impressed radial ribs on internal surface, except for ventral periphery.

Description.—Shell barely exceeding 32 mm in length and height, nearly equiconvex or only slightly right-convex in adult stages, nearly acline in young stages but tending to become slightly prosocline in adult; height subequal to length in young but becoming somewhat smaller than length in adult; thickness appearing almost isometric to length (and height) throughout growth; form ratio T/L scarcely exceeding 0.27 in both valves; hinge line moderate in length and apparently negatively allometric to overall length; test rather thick; apical angle of disk about 95 degrees; wings moderate in size, anterior one slightly larger than posterior; anterodorsal and posterodorsal margins of disk slightly concave, subequal in length in young, but with the former becoming slightly longer than the latter in adult; byssal notch moderate in depth, with relatively narrow fasciole area; ctenolium commonly indistinct, something, however, which may be due to unfavorable preservation of examined material; pre-umbonal margin of right valve highly elevated above hinge axis.

Outer surface of disk ornamented with 20–25 radial ribs, each of which is reinforced by a broad and flat-topped central solid ridge and bordered by a pair of narrow lateral sulci originally covered by oppositely disposed imbricated scales; interspaces as broad as ribs, marked with densely spaced erect scales, and lacking intercostal threads even in adult shell; sculpture of disk in both valves essentially similar; stepwise growth ring(s) sometimes observed on disk; both wings of the two valves having several (five or so) radial riblets; scales apparently undeveloped on byssal wing.

Inner surface smooth, covered by thickly deposited inner shell layer, except for ventral marginal area; cardinal crura well developed especially in right valve.

Remarks.—The present species was fully described by Masuda (1958) on the basis of abundant specimens from the Moniwa Formation in the environs of Sendai. The present sample *Mn* (*Y*) coincides well with them and also with the holotype from the Yanagawa Formation. Masuda (1958) understandably regarded this species as an early representative of *Cryptopecten*, though subsequently it was referred to *Aequipecten* together with some living species from Japan (Masuda, 1962; Masuda and Noda, 1976). Masuda (1958, 1962) noted that the number of radial ribs varies from 16 to 26 in his material, but my observation of the sample *Mn* (*Y*) and the pooled specimens in Tohoku University and the Saito Ho-on Kai Museum leads me to believe that the variation may not be so wide; most specimens possess 21 to 24 ribs. This conclusion seems to agree well with Mr. Y. Sato's unpublished morphometric data on some samples of the same species.

Owing to the coarse-grained matrix and pre-depositional abrasion, the detailed structure of the radial ornaments is in most cases difficult to observe, and imbricated scales on the hollow parts are almost entirely lost in most specimens. However, they are clearly observable in a few well-preserved specimens. The left valve illustrated by Masuda (1958, pl. 26b, fig. 8) is an exceptionally good specimen, showing the characteristic sculpture of *Cryptopecten* type and oppositely disposed imbricated scales.

Shikama (1973) recorded the occurrence of this species from the (?) Late Miocene Zushi Formation of Miura Peninsula. Though I have inquired at the store room of the Yokohama National University where the collection related to this paper is supposed to be preserved, I could not find the relevant specimens. Aside from this occurrence, hitherto known stratigraphic occurrences of *C. yanagawaensis* seem to be restricted to the Middle Miocene in north and central Honshu. Several specimens from the Akeyo Formation in the Mizunami area and the Sunakozaka Formation in the Kanazawa area are undoubtedly referable to this species.

Aequipecten matsunagiensis Masuda, 1966, from the (?) Middle Miocene Higashi-innai Formation in Ishikawa Prefecture (central Honshu), is probably not a representative of *Cryptopecten*, because the irregularly bifurcated radial ribs in the holotype (Masuda, 1966, pl. 35, fig. 4) are not to be found in this genus. I think that this species is rather closely related to *Chlamys otukae* Masuda and Sawada, 1960. However, one of the paratypes of *A. matsunagiensis* (Masuda, 1966, pl. 35, fig. 6) shows regular and simple radial ribs which remind me of those of *C. yanagawaensis*.

The present species resembles *C. bullatus* in the large number of radial ribs on the disk, comparatively weak shell convexity and low outline. In some other characteristics, especially in the considerably thick test, flat-topped central ridge of each rib and oppositely arranged imbricated scales, it is rather similar to Phenotype Q of *C. vesiculosus*. In these respects, *C. yanagawaensis* is morphologically intermediate between the two living species but clearly distinguishable from them.

Distribution.—This species is known from the Yanagawa Formation at Yanagawa, Fukushima Prefecture (type locality), the Moniwa Formation at Moniwa and Jûnishin (Masuda, 1958), Taira (Masuda and Takegawa, 1965) and Kanagase of Ôgawara (IGPS no. 90785) near Sendai, Miyagi Prefecture, the Akeyo Formation (Shukunohora Sandstone) at Mizunami and Nataki (Itoigawa, Shibata and Nishimoto, 1974), Gifu Prefecture, and the Sunakozaka Formation at Higashi-ichise near Kanazawa (Ogasawara, 1976), Ishikawa Prefecture. Early Middle Miocene (15–16 Ma).

CHAPTER 13

Summary and Conclusions

1. Systematics and Distribution

Cryptopecten Dall, Bartsch and Rehder, 1938 [type-species: *Cryptopecten alli* Dall, Bartsch and Rehder, 1938 (=*Pecten bullatus* Dautzenberg and Bavay, 1912), by original designation] is a small-sized pectinid genus, now living on sublittoral and upper bathyal sandy bottoms in the warm waters of the Indo-Pacific and western Atlantic. Fossils of this genus are considerably common in the Neogene and Quaternary of Japan. The taxonomic validity of this genus is confirmed by its unique characteristics of surface sculpture, that is, the regular and simple (neither bifurcated nor inserted) radial ribs, each of which consists of a central solid ridge and a pair of lateral hollow parts covered with numerous diaphragm-like imbricated scales. The shell is generally right-convex in the adult stage, but the highly allometric growth, the mode of which is quite different in the two valves, is also worthy of notice. *Corymbichlamys* Iredale, 1939 [type-species: *Chlamys corymbiatus* Hedley, 1909, by original designation] should be regarded as a junior synonym of *Cryptopecten*.

The following four living and two extinct species are certainly referable to *Cryptopecten*.

Pecten (*Chlamys*) *bullatus* Dautzenberg and Bavay, 1912; Philippines, Indonesia, Hawaii, South China Sea, Japan and Nasca Ridge (lower sublittoral to upper bathyal); Late Pliocene to Recent.

Pecten nux Reeve, 1853; Polynesia, Melanesia, Micronesia, eastern and northern Australia, Indonesia, Philippines, Taiwan, Japan, Andaman, Oman, Seychelles, Mauritius, Kenya, Mozambique, Madagascar, eastern South Africa and Red Sea (mainly upper sublittoral); Early Miocene to Recent.

Pecten vesiculosus Dunker, 1877; Japan (mainly lower sublittoral); Middle Pliocene to Recent.

Pecten phrygium Dall, 1886; Gulf of Mexico, east coast of the United States and north of Guiana (lower sublittoral to upper bathyal); Recent.

Pecten (*Aequipecten?*) *yanagawaensis* Nomura and Zinbo, 1936; Japan; Middle Miocene.

Cryptopecten spinosus sp. nov.; Japan; Late Pleistocene.

The shell convexity is weak in *C. bullatus* and *C. phrygium*, weak to moderate in *C. spinosus* and *C. yanagawaensis*, moderate to strong in *C. vesiculosus*, and usually very strong in *C. nux*. The number of radial ribs in the examined samples is 18–23 in *C. bullatus*, 18–22 in *C. nux*, 13–18 in *C. vesiculosus*, 17–19 in *C. phrygium*, 20–25 in *C. yanagawaensis* and 12–15 in *C. spinosus*. The imbricated scales are regularly alternate in *C. bullatus*, *C. nux* and *C. phrygium*, but always oppositely disposed in other species. The test is very thin in *C. bullatus* owing to the poorly developed inner layer, but moderate or rather thick in other species.

The following subspecies are morphologically and biogeographically distinguishable from the nominate subspecies.

Cryptopecten nux sematensis (Oyama, 1954); Japan; Late Pleistocene.

Cryptopecten vesiculosus makiyamai subsp. nov.; Japan; Late Pliocene.

C. nux sematensis is characterized by its unusually tumid outline, and *C. vesiculosus makiyamai* by its large size and exceptional occurrence in fine-grained sediments.

The specimens hitherto assigned to *"Cryptopecten tissotii"* in various Japanese reports are almost entirely referable to *C. bullatus. Cryptopecten alli* Dall, Bartsch and Rehder, is evidently a junior subjective synonym of *P. bullatus. Pecten hastingsii* Melvill, 1888, *Pecten guendolenae* Melvill, 1888, *Chlamys smithi* Sowerby, 1908, *Chlamys corymbiatus* Hedley, 1909, and *Pecten (Aequipecten) kikaiensis* Nomura and Zinbo, 1934, are certainly synonymous with *P. nux. Pecten hysginodes* Melvill, 1888, may be synonymous with *P. vesiculosus. Pecten inaequivalvis* Sowerby, 1887, and *Pecten oweni* Gregorio, 1936, which were included in *Cryptopecten* by some authors, seem to be referable to *Haumea* and *Comptopallium* (or *Decatopecten*), respectively.

2. Reproduction and Growth Rings

As the result of my observation on living populations in the eastern part of Sagami Bay carried out in several different seasons, it was found that all the large-sized individuals (over 13 mm in length and height) of *C. vesiculosus*, regardless of the phenotype, have well-developed hermaphroditic reproductive glands in summer. The curvature of shell surface, both external and internal, becomes unusually strong in this season, and a stepwise growth ring is formed. It is, therefore, reasonable to consider that the increased internal space produced by the strong curvature corresponds well to the development of gonad. Such conspicuous growth rings occur also in several other fossil and living pectinids, and, I presume, the intervals between them indicate approximate annual growth amount after the individual reached sexual maturity.

3. Continuous Variation and Relative Growth

The intrapopulational variation of various morphometric (both univariate and bivariate) and meristic characters were examined on some selected large samples of *C. vesiculosus*, *C. bullatus* and *C. nux*. Because the shell size and form ratios between various linear measurements are growth-variant and strongly influenced by the original age composition and post-mortem sorting, some standardization is necessary. In this article the partial shell height (H_1) from the origin of growth to the termination of the first growth ring, which probably signifies size at sexual maturity, was applied for the standardized shell size. The average and variability of shell shape were recognized mainly through bivariate analyses of relative growth. In almost all the samples, significant positive allometry was ascertained in the relation of convexity (T) to length (L) (only in right valve), and, on the contrary, significant negative allometry in the relation of height (H) to length (L), convexity (T) to length (L) (only in left valve) and length of two wings (D) to overall length (L). In other words, the shell outline of *Cryptopecten* generally changes from left-convex to right-convex with proportionally lowered disk and shortened wings through growth. Standardized form ratios, H/L, T/L and D/L, are obtained in each sample from the reduced major axes at several definite sizes.

The number of radial ribs, which is empirically almost growth-invariant except for very early growth stages, always shows a histogram resembling normal frequency distribution, giving a reliable basis for comparison between samples and between taxa.

4. Sculpture Dimorphism

All the Recent and post-Middle Pleistocene population samples of *C. vesiculosus* show significant dimorphism in surface sculpture; one phenotype is called Q and is characterized by the highly elevated and *quadrate* radial ribs and erect scales on the interspaces, while the other phenotype is called R and is characterized by the flattened and *rounded* radial ribs and imbricated scales covering not only the sides of ribs but also the interspaces. No individual is actually intermediate, and the relative frequency of each phenotype is clearly recognizable in all the fossil and Recent samples. No significant difference can be detected between the two phenotypes in such morphological characters as shell size, growth rate, number of ribs, allometric indices, coloration and anatomical features.

As the result of my survey on Recent samples, it was found that the two phenotypes occur in association at every locality, seeming to constitute the same interbreeding population. Such a perfect overlap of distribution with relatively stable frequency would be quite unlikely in the case of sibling species. Ecophenotypic effect and sexual dimorphism are also improbable. Though the genetic background is still unknown, various circumstantial evidences suggest that this dimorphism is controlled by a single or only a few genetic factors such as an allele or chromosomal aberration.

Somewhat similar dimorphism is found in *Aequipecten commutatus* from the Mediterranean and *Volachlamys hirasei* from Japan. The individual and geographic variations of these species should be further studied at the population level.

5. Geographic Variation

In 1973 I concluded that the geographic variation of *C. vesiculosus* in present-day seas is insignificant, so far as the phenotypic frequency is concerned. This conclusion was, however, based only on several Recent samples from the Pacific coast of central Honshu which do not cover the whole distribution of this species. As the result of my extensive survey on samples not only from the Pacific coast of Honshu but also from the East China Sea and the Japan Sea, some positive evidence of geographic variation was detected. The frequency of Phenotype R ranges from 0.39 to 0.46 in the large samples from the Pacific coast of central Honshu, whereas in the samples from the East China Sea and the Japan Sea it is much lower, ranging from 0.09 to 0.30. A similar tendency is also ascertained in the average number of radial ribs, which is as small as 14.5–14.8 in the samples from the Pacific coast of Honshu but commonly exceeds 15.0 in the samples from the East China Sea and the Japan Sea. The geographic variation of these characters seems to indicate the presence of some gentle clines along the Pacific coast of the Japanese Islands. Such clinal changes are not clearly detected on the samples along the Japan Sea side, though this may be due to the small sample sizes. The geographic variation of this species in the geologic past is still difficult to analyze, because the available fossil samples are strongly biassed in time and space.

The growth rate of shell and various bivariate characters are also often significantly different among Recent samples, but no conspicuous cline is recognized. This may be partly due to the deficiency of large samples, but, I presume, these characters may be more strongly influenced by local environmental factors than the phenotypic frequency and number of ribs.

6. Phyletic Evolution

The phyletic evolution of *C. vesiculosus* is well characterized by clear phenotypic substitution. All the Pliocene and Early Pleistocene samples are monomorphic, consisting only of the individuals of Phenotype *Q*. Phenotype *R*, so far as I am aware, first appeared in Middle Pleistocene (about 0.5 Ma) on both sides of the Japanese Islands, and the frequency seems to have increased with time. Because the amount of change evidently surpasses the range of geographic variation in the present-day seas, the phenotypic increase is mainly attributable to phyletic evolution.

The Late Pleistocene Jizôdô Formation of Boso Peninsula near Tokyo yields a number of rich fossil samples of this species. In this peninsula and its adjacent sea area the phenotypic frequency has increased from 0.15 to 0.43 during these 0.37 million years. This change could be theoretically explained by assuming a very slight difference of adaptive value (only of the order of 10^{-5} on average) between the two phenotypes. The causal evaluation of this change is at present no more than intuitive, because such non-adaptive causes as recurrent mutation and random genetic drift are also probable. It may be said, however, that this phenotypic change can be explained in any case by the acknowledged mechanism of microevolution.

The average number of radial ribs has also changed in excess of the range of geographic variation in the present seas. In most of the Late Pliocene samples the average exceeds 16.0, while it ranges from 14.5 to 15.3 in Recent samples. Several large Pleistocene samples from Boso Peninsula seem to indicate that this character has shifted directionally with time.

So far as the phenotypic frequency and the number of radial ribs are concerned, I would venture to say that the living populations off the Pacific coast of central Honshu are more advanced than those in the East China Sea and Japan Sea. The size of maximum individual and annual growth rate may be influenced by local conditions, but they are generally larger in Pliocene and Pleistocene samples than in Recent ones.

7. Speciation and Extinction

The three living Indo-Pacific species of *Cryptopecten*, i. e., *C. bullatus*, *C. nux* and *C. vesiculosus*, and probably also *C. phrygium* in the western Atlantic seem to have evolved along independent lineages at least since Late Pliocene. Although phyletic change may have occurred also in lineages other than *C. vesiculosus*, the clear morphological gaps among these lineages appear to indicate a punctuated pattern rather than gradual diversification (see Fig. 24).

Judging from the fundamental structure of radial ribs and geographic distribution, *Cryptopecten* may have already been differentiated into two species groups in Middle Miocene. One group, comprising *C. bullatus*, *C. nux* and *C. phrygium*, is characterized by the alternate arrangement of imbricated scales and extensive geographic distribution. *C. nux* is a long-ranging living species and is regarded as constituting the parental stock of this species group, because fossil records from Early Miocene onward and from

120

Middle Pliocene onward have been known in east Africa and the northwest Pacific region, respectively. *C. bullatus* was probably derived from *C. nux*, adapting now to a somewhat deeper environment in the central and northwestern Pacific. Though there is no fossil record, *C. phrygium* may have arisen by geographic isolation from a peripheral population of *C. bullatus*.

The other species group, comprising *C. vesiculosus*, *C. yanagawaensis* and *C. spinosus*, is characterized by the opposite disposition of imbricated scales and restricted geographic distribution to Japan and its adjacent seas. *C. yanagawaensis* occurs only from the Middle Miocene of Japan and may be ancestral to the post-Early Pliocene species, *C. vesiculosus*, but the wide gap of fossil records makes the interspecific relation obscure. *C. spinosus* seems to have suddenly appeared in the Late Pleistocene coral sand facies in south Japan and to have then become extinct. It shares many common characteristics with *C. vesiculosus* and probably arose from the stock of this species through some geographic speciation, though the process is as yet poorly documented.

C. nux sematensis seems to represent a Late Pleistocene local population with specialized morphology near the northern periphery of distribution of *C. nux*. This assumption is consistent with the fact that some observed samples of *C. nux nux* in the peripheral areas of the present-day distribution (e. g., Queensland and the Red Sea) also show more or less specialized morphology. *C. vesiculosus makiyamai* from the Upper Pliocene of Japan, which exceptionally occurs in fine-grained sediments, may represent an incipient branch from the stock of *C. vesiculosus*, though the possibility of ecophenotypic effect is not necessarily denied.

The restored phylogeny of *Cryptopecten* is thus characterized by the presence of a few short-ranging species and subspecies in addition to several persistent lineages. These short-ranging taxa, I believe, represent only a fraction of the speciation and peripheral isolation that actually occurred. Speciation and geographic isolation of a population (and its absorption into the parental population) are probably ubiquitous events, and the records of a large number of other unsuccessful dead end taxa have probably simply not been discovered or have been lost.

Also worthy of notice is the fact that the geographic distribution of each species of *Cryptopecten* seems to be conservative through geologic times. For example, almost all the known fossils of the three living Indo-Pacific species have been found in areas adjacent to the present-day distribution, even if the northern (and southern) limit may have fluctuated in accordance with climatic changes.

8. Significance of Phenotypic Substitution in Macroevolution

In conclusion I will summarize the significance of phenotypic substitution as one of the major features of evolution. Phenotypic substitution, here defined, is the chronological change of relative phenotypic frequency in a polymorphic species. Among neontologists this is generally recognized as transient polymorphism, typically exemplified by the industrial melanism of some moths in the 19th Century (Kettlewell, 1961; Ford, 1964). Komai's (1956) study of transient polymorphism in a ladybeetle, *Harmonia axyridis*, over a period of 40 years suggested that some selection pressure caused by long-term climatic change is responsible for the observed definite trend. These are, of course, excellent examples of microevolution, but probably the longest observations on actual living populations.

The phyletic evolution of *C. vesiculosus* seems to involve phenotypic substitution over the last 500,000 years. Though the causal reasoning of this change, whether it is adaptive or non-adaptive, may be difficult to establish, this phenomenon suggests that studies of transient polymorphism could be expanded to the geological past. Polymorphic relation between two or more nominal species and sometimes phenotypic substitution have also been postulated in some fossil animals; Pleistocene-Recent bears, *Ursus*, by Kurtén (1955), Pliocene sand dollars, *Merriamaster*, by Durham (1978), and Cretaceous ammonites, *Gaudryceras*, by Hirano (1978), for example. Although I am not necessarily in a position to confirm the adequacy of their interpretations, with careful studies of the relation between phena and taxa, phenotypic substitution could be recognized as ubiquitous phenomena in various taxonomic groups.

Phenotypic substitution seems to be one of the most intrinsic causes for the punctuated patterns of morphological change in the fossil records. Problems about its role in macro-evolution may be problems of scale—how significant morphological change actually occurs by a single mutation. The morphological discontinuity between the two phenotypes of *C. vesiculosus* appears to be slight in one aspect but considerable in another. The two phenotypes are not significantly different in numerous unit characters, but the surface sculpture of the mutant (Phenotype *R*) is quite new and unique in the genus *Cryptopecten* and probably also in the family Pectinidae. Hollow structure of radial ribs is regarded as an important diagnostic character of *Cryptopecten*, but, in one and the same population of *Aequipecten commutatus*, hollow structure may or may not be present. As has been discussed by many evolutionists (e. g., Mayr, 1963), instantaneous speciation may be almost impossible in many animal groups in which polyploidy is rare, but, I believe, the ultimate origin of a new taxonomic character, even for a taxon higher than species, should sometimes be sought in a single gene (especially "regulatory gene") and chromosome mutation, and subsequent phenotypic substitution.

9. Further Problems

The final object of this study is to understand the evolutionary pattern and process of a particular taxonomic group, a pectinid genus in this case, from various paleontological and neontological viewpoints. As the first step, the morphological change in a part of the constituent species has been clarified to some extent at the population level, but many problems remain unsolved. With respect to *C. vesiculosus*, intrapopulational and geographic variations as well as chronological shift of shell morphology were detected for several characters; as the next step, developmental, cytological and genetical examinations are needed, particularly in relation to the background of the observed dimorphism. For other species of *Cryptopecten*, present knowledge is still far from what is required for completion of the first step, and effort should be devoted to obtaining basic information on distribution, ecology and morphology. More extensive surveys on living and fossil populations of these species are of course indispensable to the task of reconstructing their evolutionary process and interspecific relationships. The present study has brought home to me once again the profundity of the study of evolution and convinced me anew that the best understanding of evolutionary process will be accomplished by a synthesis of the results obtained through various approaches, each with their advantages and limitations.

List of Examined Samples

Repository: Department of Historical Geology and Palaeontology, University Museum, University of Tokyo, unless otherwise stated

N: Number of total individuals

N_R: Number of individuals belonging to Phenotype R (only for *C. vesiculosus*)

N_Q: Number of individuals belonging to Phenotype Q (only for *C. vesiculosus*)

CV: Number of conjoined valves (Recent specimens mostly alive)

RV: Number of right odd valves

LV: Number of left odd valves

Cryptopecten bullatus
(Dautzenberg and Bavay)
Fossil sample

Sample *Kk* (B) [UMUT CM16001]

Locality: Road-cut about 500m north of Kamikatetsu, Kikai Island (Kikai Town), Kagoshima Prefecture [28°17.0′N, 129°56.8′E]

Horizon: Coral sand of the Wan Formation of Ryukyu Group

Age: Late Pleistocene (N23, 0.08 Ma, according to Sakanoue et al., 1967, and Omura, 1983)

Collector: Yamaguchi and Hayami

Composition: $N=62$, RV$=36$, LV$=26$

Recent samples

Sample *Ma* (B) [UMUT RM16009]

Locality: about 5 km west from the western end of Jôgashima Islet, Sagami Bay [35°08.5′N, 139″33.7′E, 105m]

Collector: Hayami et al. (R/V Rinkai)

Composition: $N=1$, LV$=1$

Sample *Hs* (B)

Locality: Hyôtanse, a sea bank west of Niijima Island (3 stations)

Collector: Okutani (R/V Soyo II)

Composition (integrated): $N=12$, CV$=3$, RV$=4$, LV$=5$

[Subsample *Hs* (*3*) (B)] [UMUT RM16002] (D15 by Okutani, 1972, 34°21.0′N, 139°03.8′E, 170 m) $N=6$, CV$=2$, RV$=1$, LV$=3$

[Subsample *Hs* (*8*) (B)] [UMUT RM16003] (D55 by Okutani, 1972, 34°21.9′N, 139°05.7′E, 220 m) $N=5$, RV$=3$, LV$=2$

[Subsample *Hs* (*9*) (B)] [UMUT RM16004] (D29 by Okutani, 1972, 34°21.5′N, 139°04.5′E, 200 m) $N=1$, CV$=1$

Sample *Km* (B)

Locality: Shelf off Kushimoto of Kii Peninsula (2 stations)

Collector: Tsuchida

Composition (integrated): $N=2$, RV$=1$, LV$=1$

[Subsample *Km* (*1*) (B)] [UMUT RM16005] (off Kushimoto, 110–115 m) $N=1$, LV$=1$

[Subsample *Km* (*2*) (B)] [UMUT RM16006] (off Kushimoto, 80 m) $N=1$, RV$=1$

Sample *Sr* (B)

Locality: off Setozaki, Shirahama of Kii Peninsula (3 stations)

Collector: Yamaguchi, Oji and Kamiya

Composition (integrated): $N=28$, CV$=11$, RV$=7$, LV$=10$

[Subsample *Sr* (*1*) (B)] [UMUT RM16183] (ca. 6 km SSW of Setozaki, 33°36.0′N, 135°18.0′E, 140 m) $N=1$, CV$=1$

[Subsample *Sr* (*2*) (B)] [UMUT RM16184] (same as above, 140–160 m) $N=22$, CV$=7$, RV$=5$, LV$=10$

[Subsample *Sr* (*3*) (B)] [UMUT RM16185] (same as above, 180 m) $N=5$, CV$=3$, RV$=2$

Sample *Ks* (B) [UMUT RM16007]

Locality: Mouth of Kii Strait, 300 m (unknown in detail)

Collector: Hayashi (through fishery)

Composition: $N=2$, CV$=2$

Sample *Kj* (B) [UMUT RM16008]

Locality: off Koshikijima Island, west of Kyushu (Loc. 296, 32°49.5′N, 128°07.5′E)

Collector: Oyama (R/V Tokai II)

Composition: $N=2$, LV$=2$

Sample *Sk* (B) [UMUT RM16181]

Locality: northeast of Senkaku Islets, East China Sea (KH68–2–T–12, 26°55′N, 125°01′E, 120 m)

Collector: Tsuchida (R/V Hakuho)

Composition: $N=1$, CV$=1$

Sample *Ta* (B)

Locality: Tsushima Strait, southeast of Shimoagata Island (St. 8 by Imajima, 1970,

33°49.7′N, 129°28.9′E, 100 m)
Collector: Habe and Imajima (R/V Genkai)
Composition: $N=6$, RV$=3$, LV$=3$
Repository: Department of Zoology, National Science Museum, Tokyo
Sample *Bh* (*B*) [UMUT RM16195]
  Locality: the sea around Bohol Island, Philippines (unknown in detail)
  Collector: Kase through a shell collector at Bohol
  Composition: $N=1$, CV$=1$

Other confirmed Recent specimens of *C. bullatus**

USNM 173194. Off south coast of Oahu, Hawaii (238–252 fms, USBF station 3811); Holotype of *Cryptopecten alli*. $N=1$, CV$=1$

USNM 190440. Vicinity of Kauai Island, Hawaii (257–312 fms). $N=1$, LV$=1$

USNM 335667. West coast of Island of Hawaii (198–147 fms). $N=1$, LV$=1$

USNM 764151. Off Waikiki, Oahu, Hawaii (190 fms). $N=5$, RV$=2$, LV$=3$

USNM 204163. Off Nagasaki, Kyushu, Japan (107 fms, USBF Station 4904). $N=1$, RV$=1$

USNM 204165. Kagoshima Gulf, Japan (103 fms, USBF Station 4936). $N=2$, RV$=1$, LV$=1$

USNM 204277. "Japan Seas" (106 fms). USBF Station 4893. $N=2$, RV$=2$

USNM 254287. Batangas Bay, Luzon, Philippines (170 fms, USBF Station 5268). $N=1$, RV$=1$

USNM 294674. Ragay Gulf, off N. Burias, Philippines (105 fms, USBF Station 5217). $N=2$, RV$=1$, LV$=1$

USNM 295635. Off Eascarceo Point, N. Mindoro, Philippines (244 fms, USBF Station 5294). $N=1$, LV$=1$

USNM 295872. Off Matocot Point, W. Luzon, Philippines (170 fms, USBF Station 5268). $N=2$, RV$=2$

USNM 297072. South China Sea, off Pratas Island (230 fms, USBF Station 5317). $N=3$, RV$=3$

USNM 298318. Off Gigantangan Island, northwest Leyte, Philippines (114 fms, USBF Sta-

---

* Because the present collection of *C. bullatus* does not cover the distribution of this species, the specimens in the National Museum of Natural History, Washington D. C. and the American Museum of Natural History, New York were also used in the present study. Locality and other information (if any) has been taken directly from the museum labels. The USBF stations are from the cruises of the R/V Albatross (U.S. Bureau of Fisheries).

tion 5398). $N=1$, LV$=1$

USNM 764152. Gulf of Tosa, Japan (100 fms). $N=8$, CV$=8$

USNM 773983. Nasca Ridge (ca. 750 miles west of north Chile) (Downwind Expedition Station 73, 25°44′S, 85°25′W, 228 m). $N=3$, RV$=3$

AMNH 127494. Tosa, Japan (100 fms). $N=2$, CV$=2$

AMNH 171523. Tosa, Japan (70 fms). $N=4$, CV$=4$

AMNH 183928. Tosa, Japan. $N=1$, CV$=1$

AMNH 127507. Kii, Japan (70 fms). $N=1$, CV$=1$

Cryptopecten nux (Reeve)
Fossil samples

Sample *Mk* (*N*) [UMUT CM16010]
  Locality: Yonahama, east coast of Miyako Island, Okinawa Prefecture [24°47.1′N, 125°22.4′E]
  Horizon: Medium sandstone of Yonahama Formation of Shimajiri Group
  Age: Middle Pliocene (N20–N21, ca. 3.0 Ma)
  Collector: Ozawa, Yamaguchi and Hayami
  Composition: $N=10$, RV$=5$, LV$=5$
Sample *Ik*(*N*) [UMUT CM16011]
  Locality: Northern coast of Ikeijima Islet, Yonagusuku-son, Okinawa Island, Okinawa Prefecture [26°23.8′N, 128°00.0′E]
  Horizon: Medium-coarse sandstone of Shinzato Formation of Shimajiri Group
  Age: Late Pliocene (N21, 2.0–2.5 Ma)
  Collector: Yabu, Yamaguchi, Tabuki and Hayami
  Composition: $N=4$, RV$=1$, LV$=3$
Sample *Ob* (*N*) [UMUT CM16012]
  Locality: Shimo-oyakebaru, Sahiki-son, Okinawa Island, Okinawa Prefecture [26°09.7′N, 127°46.9′E]
  Horizon: Medium sand of Chinen Formation of Shimajiri Group
  Age: Early Pleistocene (N22, 1.7–2.0 Ma)
  Collector: Nohara
  Composition: $N=5$, RV$=3$, LV$=2$
Sample *Mt* (*N*) [UMUT CM16013]
  Locality: Northwest of the Station Tsukuihama of Keihin Express Line, Nakakôji, Tsukui, Yokosuka City, Kanagawa Prefecture [35°11.8′N, 139°40.1′E]
  Horizon: Medium sand of the Miyata Formation
  Age: Middle Pleistocene (unknown in detail)
  Collector: Kase
  Composition: $N=1$, RV$=1$

124

Sample *Jz* (*N*) [UMUT CM16014]
Locality: Road-cut, east of a shrine, Jizôdô, Kisarazu City, Chiba Prefecture [35°22.0′N, 140°05.9′E]
Horizon: Coarse sand of Jizôdô Formation of Shimosa Group
Age: Late Pleistocene (N23, ca. 0.37 Ma)
Collector: Hayami
Composition: *N*=1, RV=1

Sample *Ic* (*N*) [UMUT CM16029]
Locality: Road-cut, east of Ichinosawa, Kisarazu City, Chiba Prefecture [35°20.8′N, 140°04.2′E]
Horizon: Medium sand of Jizôdô Formation of Shimosa Group
Age: Late Pleistocene (N23, ca. 0.37 Ma)
Collector: Ohta
Composition: *N*=1, LV=1

Sample *Sm* (*N*) [UMUT CM16180] [*Cryptopecten nux sematensis* (Oyama)]
Locality: Lower part of a cliff of Semata-no-seki, Ochishimoshinden, Toki Town, Chiba Prefecture [35°31.5′N, 140°13.7′E]
Horizon: Medium sand of Yabu Formation (layer E of Ohara, 1968a) of Shimosa Group
Age: Late Pleistocene (N23, ca. 0.29 Ma)
Collector: Ohara, Akutsu and Hayami
Composition: *N*=12, RV=6, LV=6
Remarks: Two specimens described by Yokoyama (1922) [UMUT CM21561, 21562] probably occurred at the same locality.

Sample *Kk* (*N*) [UMUT CM16015]
Locality: Road-cut about 500 m north of Kamikatetsu, Kikai Island (Kikai Town), Kagoshima Prefecture [28°17.0′N, 129°56.8′E]
Horizon: Coral sand of Wan Formation of Ryukyu Group
Age: Late Pleistocene (N23, ca. 0.08 Ma, according to Sakanoue et al., 1967, and Omura, 1983]
Collector: Yamaguchi and Hayami
Composition: *N*=1682, CV=1, RV=866, LV=815

Other confirmed fossil specimens of *C. nux*
IGPS no. 50357. Ryukyu Limestone at plateau near Kamikatetsu, Kikai Island, Kagoshima Prefecture. (type material of *Pecten* (*Aequipecten*) *kikaiensis* described by Nomura and Zinbo (1934))
IGPS no. 13369. [*Cryptopecten nux sematensis* (Oyama)] "Narita Bed at Takakura, Shitô Village, Chiba Prefecture" (pars) (collected by Yabe)

Recent samples

Sample *Zs* (*N*)
Locality: Zenisu, a sea bank southwest of Kôzushima Island (2 stations)
Collector: Okutani (R/V Soyo II)
Composition (integrated): *N*=3, RV=3
[Subsample *Zs* (*8*) (*N*)] [UMUT RM16016] (D50 by Okutani, 1972, 34°00.6′N, 138°50.2′E, 105 m) *N*=2, RV=2
[Subsample *Zs* (*9*) (*N*)] [UMUT RM16017] (unknown in detail) *N*=1, RV=1

Sample *Ts* (*N*) [UMUT RM16018]
Locality: Takase, a sea bank northwest of Niijima Island [D58 by Okutani, 1972, 34°26.7′N, 130°11.3′E, 95 m]
Collector: Okutani (R/V Soyo II)
Composition: *N*=2, LV=2

Sample *Su* (*N*) [UMUT RM16019]
Locality: Senoumi, a sea bank in Suruga Bay [St. 15 by Sanyo Suiro Co., 34°44′28″N, 138°31′19″E, 56 m] (the same locality as sample *Su* (*10*))
Collector: Sanyo Suiro Co.
Composition: *N*=2, RV=1, LV=1

Sample *An* (*N*) [UMUT RM16020]
Locality: Mouth of Ise Bay, off Anori-Matoya [30 m]
Collector: Hayashi (through fishery)
Composition: *N*=1, CV=1

Sample *Km* (*N*)
Locality: Shelf off Kushimoto of Kii Peninsula (4 stations)
Collector: Tsuchida
Composition (integrated): *N*=10, RV=7, LV=3
[Subsample *Km* (*1*) (*N*)] [UMUT RM16021] (off Kushimoto, 110–115 m) *N*=1, RV=1
[Subsample *Km* (*2*) (*N*)] [UMUT RM16022] (off Kushimoto, 80 m) *N*=1, LV=1
[Subsample *Km* (*12*) (*N*)] [UMUT RM16023] (off Kushimoto, 32 m) *N*=2, RV=1, LV=1
[Subsample *Km* (*13*) (*N*)] [UMUT RM16024] (off Kushimoto, 60–80 m) *N*=6, RV=5, LV=1

Sample *Bh* (*N*) [UMUT RM16194]
Locality: off Panglao, Bohol Island, Philippines (unknown in detail)
Collector: Kase through a shell collector at Bohol
Composition: *N*=19, CV=19

Other confirmed Recent specimens of *C. nux**

---

* Because the present collection of *C. nux* does not cover the distribution of this species, the

USNM 718977. Faqhuar Group of Seychelles (9°36′N, 51°01′E, 80 m)

USNM 763466. Ditto (10°00′N, 51°15′E, 60 m)

USNM 719012. Ditto (9°41′N, 51°03′E)

USNM 718636. 80 miles east of Beira, Mozambique (19°51′S, 36°21′E, 62 m)

USNM 718625. Ditto (19°50′S, 36°21′E, 65 m)

USNM 717132. 60 miles northeast of Durban, South Africa (29°11′S, 32°02′E)

USNM 763465. 35 miles west of Cape St. Andre, Madagascar (16°11′S, 43°42′E)

USNM 763464. 65 miles southwest of Cape St. Andre, Madagascar (16°42′S, 43°19′E, 150–300 m)

USNM 763463. 50 miles southeast of Beira, Mozambique (20°30′S, 35°43′E, 62 m)

USNM 763462. 40 miles northeast of Durban, South Africa (29°19′S, 32°00′E, 366 m)

USNM 763459. 160 miles southeast of Rodriques Island off Mauritius (21°21′S, 65°52′E)

USNM 763460. 80 miles south-southwest of Pontada Barra, Mozambique (24°49′S, 35°13′E, 73 m)

USNM 764165. Al Ghardaqa, Egypt (Red Sea) (242 m)

USNM 762403. Lourence Marques, Indian Ocean (42 m)

USNM 764161. Andaman Island, India

USNM 764163. Amirante Island, southwest of Seychelles (34 fms)

USNM 761546a. 40 miles east of Quissico, Mozambique (110 m)

USNM 764203. [coast of south Mozambique] (24°53′S, 34°56′E, 55 m)

USNM 674222. [Ditto] (24°48′S, 34°59′E, 42 m)

USNM 348647. Cairns Reef, north Queensland, Australia (6–10 fms)

USNM 764159. Off Hope Island, Queensland, Australia (5–10 fms). Topotype of *Chlamys corymbiatus*.

USNM 764157. Hervey Bay, Queensland, Australia (30 fms)

USNM 764156. Cannonvale Beach, Whitsunday Passage, Queensland, Australia

USNM 764158. Singapore Island

USNM 299742, 299775, 299814. Off Sipadan

Island, Sibuko B., Borneo (USBF Station 5586, 347 fms)

USNM 230228, 230314. Off Jolo Jolo, Philippines (USBF Station 5137, 20 fms)

USNM 235461. Ditto (USBF Station 5136, 22 fms)

USNM 235488. Northwest of Jolo Jolo, Philippines (USBF Station 5138, 19 fms)

USNM 235515. North off Jolo Jolo, Philippines (USBF Station 5139, 20 fms)

USNM 235589, 235596. Off Jolo Jolo, Philippines (USBF Station 5141, 29 fms)

USNM 235658. Northeast of Jolo Jolo, Philippines (USBF Station 5144, 19 fms)

USNM 235666, 235702, 235727, 235800, 235802. Off Jolo Jolo, Philippines (USBF Station 5145, 23 fms)

USNM 235834. Off Sulade Island, Tapul, Philippines (USBF Station 5146, 24 fms)

USNM 235989, 236004, 236005. Off Cacataau Island, Tawi Tawi, Philippines (USBF Station 5151, 24 fms)

USNM 236013. Off Tambagaau Island, Tawi Tawi, Philippines (USBF Station 5152, 34 fms)

USNM 236225. Off Bakun R. Tawi Tawi, Philippines (USBF Station 5159, 10 fms)

USNM 236566. Off Jolo Jolo, Philippines (USBF Station 5174, 20 fms)

USNM 236617, 236784. Off northeast of Tablas, Philippines (USBF Station 5179, 37 fms)

USNM 237035. Off northeast Burias, Philippines (USBF Station 5218, 20 fms)

USNM 237118, 237119. Davao Gulf, off Lanang Point, Philippines (USBF Station 5249, 23 fms)

USNM 237177. Davao Gulf, off Linao Point, Philippines (USBF Station 5251, 20 fms)

USNM 237210. Davao Gulf, off Linao Point, Philippines (USBF Station 5252, 28 fms)

USNM 237235. Pakiputau Strait, Mindanao, Philippines (USBF Station 5253, 28 fms)

USNM 237284, 237302, 237329. Davao Gulf, off Linao Point, Philippines (USBF Station 5254, 21 fms)

USNM 237384. Davao Gulf, off Dumalog Island, Philippines (USBF Station 5255, 100 fms)

USNM 237978. 17 miles northeast of Balabac Strait, Philippines (USBF Station 5355, 44 fms)

USNM 246693. Off Tataau Islands, Philippines (USBF Station 5154, 12 fms)

USNM 254209. Port Busin, Burias, Philippines

USNM 254284, 254285, 254286. South of Pangasinan, Philippines (USBF Station 5138, 19 fms)

USNM 254292. Gulf of Davao, Mindanao, Philip-

observed specimens in the National Museum of National History, Washington D. C. are included in the hypodigm. Locality and other information (if any) has been taken directly from the museum labels. The USBF stations are from the cruises of the R/V Albatross (U.S. Bureau of Fisheries).

pines (USBF Station 5253, 28 fms)

USNM 254932, 254935. Pakiputan Strait, Philippines (USBF Station 5250, 23 fms)

USNM 257745, 257859. Off Tinakta Island, Tawi Tawi, Philippines (USBF Station 5161, 16 fms)

USNM 259883. Off Matacot Point, west Luzon, Philippines (USBF Station 5296, 210 fms)

USNM 283422. Off southeast Tawi Tawi, Philippines (USBF Station 5164, 18 fms)

USNM 292042. Off Sirun Island, Tawi Tawi, Philippines (USBF Station 5151, 24 fms)

USNM 292135. Off Tinakta Island, Tawi Tawi, Philippines (USBF Station 5156, 18 fms)

USNM 292230. Off Tinakta Island, Tawi Tawi, Philippines (USBF Station 5162, 230 fms)

USNM 292358, 292359. Off Point Origon, Tablas, Philippines (USBF Station 5178, 73 fms)

USNM 292450. Off Sulade Island, Sulu Archipelago, Philippines (USBF Station 5146, 24 fms)

USNM 292537. Off Sirun Island, Tawi Tawi, Philippines (USBF Station 5148, 17 fms)

USNM 293010. Off Tocanhi Point, Tawi Tawi, Philippines (USBF Station 5153, 49 fms)

USNM 293128. Southeast off Bantayan Island, Philippines (USBF Station 5192, 32 fms)

USNM 293169. Sogod Bay, Leyte, Philippines (USBF Station 5202, 502 fms)

USNM 293781. Off Baluk-Baluk Island, Sulu Archipelago, Philippines (USBF Station 5134, 25 fms)

USNM 293815. Off Jolo Jolo Island, Philippines (USBF Station 5140, 20 fms)

USNM 293989, 294004. Off Jolo Jolo Island, Philippines (USBF Station 5142, 21 fms)

USNM 294065. Off Jolo Jolo Island, Philippines (USBF Station 5143, 19 fms)

USNM 294099. Off Jolo Jolo Island, Philippines (USBF Station 5141, 29 fms)

USNM 294215. Off Corregidor, Philippines (USBF Station 5106, 37 fms)

USNM 294287. Off Jolo Jolo Island, Philippines (USBF Station 5144, 19 fms)

USNM 294479, 294503. Off Jolo Jolo Island, Philippines (USBF Station 5139, 20 fms)

USNM 294580, 294581. Ditto (USBF Station 5136, 22 fms)

USNM 295048, 295073, 295126. Off Tinakta Island, Tawi Tawi, Philippines (USBF Station 5159, 10 fms)

USNM 295318. Off Davao, Mindanao, Philippines (USBF Station 5255, 100 fms)

USNM 295343. Off Bongo Island, Illana Bay, Mindanao, Philippines (USBF Station 5257, 28 fms)

USNM 295750. Off Matocot Point, west Luzon, Philippines (USBF Station 5269, 220 fms)

USNM 295885. Ditto (USBF Station 5268, 170 fms)

USNM 296144. Off Malavatuan Island, west Luzon, Philippines (USBF Station 5276, 18 fms)

USNM 296195. Ragay Gulf, off north Burias, Philippines (USBF Station 5218, 20 fms)

USNM 297577, 297605. Northeast off Balabac, Balabac Island, Philippines (USBF Station 5357, 68 fms)

USNM 298548. Off Cabugan Grande Island, east Leyte, Philippines (USBF Station 5481, 61 fms)

USNM 299072. Off Dammi Island, Sulu Archipelago, Philippines (USBF Station 5565, 243 fms)

USNM 299136. Off Simaluc Island, Tawi Tawi, Philippines (USBF Station 5569, 303 fms)

USNM 299259. Ditto. (USBF Station 5571, 340 fms)

USNM 299404, 299446. North off Tawi Tawi, Philippines (USBF Station 5577, 240 fms)

USNM 300408. Off Lauis Point, east Cebu, Philippines (USBF Station 5417, 165 fms)

USNM 302288. Off Pajumajan Island, Tawi Tawi, Philippines (USBF Station 5152, 34 fms)

USNM 311883, 311886, 312104, 312109, 312112. Ragay Gulf off north Burias, Philippines (USBF Station 5218, 20 fms)

USNM 431209. Off Jolo Jolo Island, Philippines (USBF Station 5138, 19 fms)

USNM 704253. Off Observatory Island, Linapacan Strait, Philippines (USBF Station 5334, 46 fms)

USNM 704252. Southeast off Bantayan Island, Philippines (USBF Station 5192, 32 fms)

USNM 704251. Off Baluk-Baluk Island, Philippines (USBF Station 5134)

USNM 764164. Jolo, Sulu Archipelago, Philippines (25 fms)

USNM 764162. Sacol, Mindanao, Philippines (65 m)

USNM 704240. Off Tinakta Island, Tawi Tawi, Philippines (USBF Station 5161, 16 fms)

USNM 237707, 704254. Off Pratas Island, South China Sea (USBF Station 5311, 88 fms)

USNM 764160. Itoman, Okinawa, Japan (50 fms)

USNM 764155. Macclesfield Bank, South China Sea (40–45 fms)

USNM 774213. Off southeast Tawi Tawi, Philippines (USBF Station 5164, 18 fms)

USNM 774214. Off Baluk-Baluk Island, Sulu Archipelago, Philippines (USBF Station 5134,

25 (fms)

USNM 774215. Off Jolo Jolo, Philippines (USBF Station 5144, 19 fms)

USNM 774216. Off Pratas Island, South China Sea (USBF Station 5314, 122 fms)

USNM 630953. Rabaul, New Britain

USNM 807368. Bramble Island, Fly River, Papua

USNM 598694. Seaward (N) of western end of Bikini Island (11°38′33″N, 165°31′18″E)

USNM 764166. Bay of Traitors, Hivaoa Island, Marquesas Islands. Topotype of *C. nux*.

USNM 798270, 798271, 794201, 794210. South of Baie de Vaieo, west coast of Ua Pou, Marquesas Islands

Specimens in M. Horikoshi's private collection. Cebu, Philippines

Ditto. Bohol, Philippines (ca. 50 m)

Ditto. Rabaul, New Britain

Cryptopecten vesiculosus (Dunker)
Fossil samples

Sample *Sh 1* [UMUT CM16025]

Locality: Coast at the north of Shirahama Shrine, Shirahama, Shimoda City, Shizuoka Prefecture [34°41.5′N, 138°58.6′E]

Horizon: Medium-coarse sandstone of Harada Formation of Shirahama Group

Age: Middle Pliocene (N19, ca. 3.5 Ma)

Collector: Hirano and Hayami

Composition: $N=229$, $N_R=0$, $N_Q=229$, RV $=101$, LV$=128$

Sample *Sh 2* [UMUT CM16026]

Locality: Outcrop at beach of Nagata, Shirahama, Shimoda City, Shizuoka Prefecture [34°41.7′N, 138°58.6′E]

Horizon: Conglomeratic coarse sandstone of Harada Formation of Shirahama Group

Age: Middle Pliocene (N19, ca. 3.5 Ma)

Collector: Hirano and Hayami

Composition: $N=16$, $N_R=0$, $N_Q=16$, RV$=7$, LV$=9$

Sample *Ht* [UMUT CM16027]

Locality: Southern coast of Cape Tsurukubi, Miyata, Hitachi City, Ibaraki Prefecture [36°35.7′N, 140°34.9′E]

Horizon: Coarse sandstone of Hitachi Formation of Taga Group

Age: Pliocene (unknown in datail)

Collector: Hayami

Composition: $N=7$, $N_R=0$, $N_Q=7$, RV$=5$, LV$=2$

Sample *Iy* [UMUT CM16028]

Locality: South of a pond, Ishiyama, Amaha

Town, Chiba Prefecture (Loc. 2 by Ohara and Takahashi, 1975) [35°11.8′N, 139°57.4′E]

Horizon: Coarse sandstone of Kurotaki Formation of Kazusa Group

Age: Late Pliocene (N21, 2.0–3.0 Ma)

Collector: Yamaguchi and Hayami

Composition: $N=21$, $N_R=0$, $N_Q=21$, RV$=11$, LV$=10$

Sample *Nj*

Locality: Chôjagakubo, Kamigô-chô, Totsuka-ku, Yokohama City (Loc. 326 by Shikama and Masujima, 1969) [35°20.3′N, 139°35.1′E]

Horizon: Medium sandstone of Nojima Formation of Kazusa Group

Age: Late Pliocene (unknown in detail)

Collector: Masujima

Composition: $N=29$, $N_R=0$, $N_Q=29$, RV$=16$, LV$=13$

Repository: Geological Institute, Faculty of Education, Yokohama National University

Sample *Ik* [UMUT CM16030]

Locality: Northern coast of Ikeijima Islet, Yonagusuku-son, Okinawa Island, Okinawa Prefecture [26°24.8′N, 128°00.0′E]

Horizon: Medium-coarse sandstone of Shinzato Formation of Shimajiri Group

Age: Late Pliocene (N21, 2.0–2.5 Ma)

Collector: Yabu, Yamaguchi, Tabuki and Hayami

Composition: $N=154$, $N_R=0$, $N_Q=154$, CV$=8$, RV$=73$, LV$=73$

Sample *Sw* [UMUT CM16031]

Locality: Tanaka, Sawada Town, Sado Island, Niigata Prefecture [38°00.8′N, 138°16.7′E]

Horizon: Medium sandstone of Sawane Formation

Age: Pliocene? (unknown in detail)

Composition: $N=3$, $N_R=0$, $N_Q=3$, RV$=1$, LV$=2$

Sample *Ne* [UMUT CM16032]

Locality: Quarry at Byôbudani, west of Naoetsu, Jôetsu City, Niigata Prefecture [37°09.8′N, 138°13.0′E]

Horizon: Coarse sand of "Byôbudani Formation"

Age: Pliocene? (unknown in detail)

Collector: Yamaguchi and Hayami

Composition: $N=32$, $N_R=0$, $N_Q=32$, RV$=14$, LV$=18$

Sample *Kg 1* [UMUT CM16033] [*C. vesiculosus makiyamai* sp. nov.]

Locality: West of Ugari, northeast of Yamanashi, Fukuroi City, Shizuoka Prefecture [34°47.5′N, 137°56.3′E]

Horizon: Fine silt of Hosoya Silt of Kakegawa

128

Group

Age: Late Pliocene (N22, 1.9–2.4 Ma)

Collector: Matsukuma and Hayami

Composition: $N=20$, $N_R=0$, $N_Q=20$, RV= 6, LV=14

Sample *Kg 2* [UMUT CM16034] [*C. vesiculosus makiyamai* subsp. nov.]

Locality: West of Ugari, northeast of Yamanashi, Fukuroi City, Shizuoka Prefecture [34°47.6′N, 137°55.4′E]

Horizon: Fine silt of Hosoya Silt of Kakegawa Group

Age: Late Pliocene (N22, 1.9–2.4 Ma)

Collector: Hayami

Composition: $N=6$, $N_R=0$, $N_Q=6$, RV=4, LV=2

Sample *Mz 1*

Locality: Hagenoshita, west of Takanabe, Takanabe Town, Miyazaki Prefecture [32°08.6′N, 131°30.9′E] (same locality as MI-5528 by Shuto (1961))

Horizon: Fine sand of Takanabe Formation of Miyazaki Group

Age: Late Pliocene (unknown in detail)

Collector: Hayami

Composition $N=5$, $N_R=0$, $N_Q=5$, RV=2, LV=3

Remarks: This sample was used by Hayami (1973) but is now missing.

Sample *Mz 2* [UMUT CM16035]

Locality: Outcrop near JNR railway bridge, south of Tôriyama-hama, Kawaminami Town, Miyazaki Prefecture [32°09.8′N, 131°32.9′E]

Horizon: Medium sand of Tôriyama Formation of Miyazaki Group

Age: Late Pliocene (unknown in detail)

Collector: Ozawa

Composition: $N=13$, $N_R=0$, $N_Q=13$, RV=7, LV=6

Sample *Ob* [UMUT CM16036]

Locality: Shimo-oyakebaru, Sashiki-son, Okinawa Island, Okinawa Prefecture [26°09.7′N, 127°46.9′E]

Horizon: Medium sand of Chinen Formation of Shimajiri Group

Age: Early Pleistocene (N22, 1.7–2.0 Ma)

Collector: Nohara

Composition: $N=85$, $N_R=0$, $N_Q=85$, RV= 46, LV=39

Sample *Oe* [UMUT CM16037]

Locality: West coast of Okidomari, east of Cape Tamina, Okinoerabu Island, Kagoshima Prefecture [27°23.7′N, 128°33.2′E]

Horizon: Coarse sand of Shimoshiro Formation of Ryukyu Group

Age: Pleistocene (unknown in detail)

Collector: Fujii

Composition: $N=2$, $N_R=0$, $N_Q=2$, RV=1, LV=1

Sample *Tm 1*

Locality: Road-cut, east of Tsujimori, Seiwa Village, Chiba Prefecture [35°13.4′N, 140°01.9′E] (The indication of the locality of this sample (*Tm*) in Hayami (1973, p. 403) is erroneous.)

Horizon: Coarse sand of Umegase (or Sakabata) Formation of Kazusa Group

Age: Middle Pleistocene (N22, 0.7–1.0 Ma)

Collector: Matsushima

Composition: $N=36$, $N_R=0$, $N_Q=36$, RV= 20, LV=16

Repository: Geological Institute, Faculty of Education, Yokohama National University

Sample *Tm 2* [UMUT CM16038]

Locality, horizon and age: Same as sample *Tm 1*

Collector: Yamaguchi and Hayami

Composition: $N=11$, $N_R=0$, $N_Q=11$, RV=6, LV=5

Sample *Ij 1* [UMUT CM16039]

Locality: Quarry at Shibahara, north of Ichijuku, Koito Town, Chiba Prefecture [35°16.2′N, 139°59.1′E]

Horizon: Coarse sand of Ichijuku Formation of Kazusa Group

Age: Middle Pleistocene (ca. 0.6 Ma)

Collector: Ohara

Composition: $N=25$, $N_R=0$, $N_Q=25$, RV= 12, LV=13

Sample *Ij 2* [UMUT CM16040]

Locality horizon and age: Same as sample *Ij 1*

Collector: Hayami

Composition: $N=8$, $N_R=0$, $N_Q=8$, RV=4, LV=4

Sample *Mt 1* [UMUT CM16041]

Locality: Northwest of Tsukuihama Station of Keihin Express Line, Nakakôji, Tsukui, Yokosuka City, Kanagawa Prefecture [35°11.8′N, 139°40.1′E]

Horizon: Medium sand of Miyata Formation

Age: Middle Pleistocene (unknown in detail)

Collector: Kase

Composition: $N=10$, $N_R=0$, $N_Q=10$, RV= 4, LV=6

Sample *Mt 2* [UMUT CM16187]

Locality, Horizon and age: Same as sample *Mt 1*

Collector: Ohta

Composition: $N=5$, $N_R=0$, $N_Q=5$, RV$=2$, LV$=2$

Sample *Nk* [UMUT CM16188]
Locality: About 500 m north of Furuyado, Shizuoka City [34°57.5′N, 138°27.4′E]
Horizon: Silty sand of Nekoya Formation
Age: Middle Pleistocene (unknown in detail)
Collector: Kondo
Composition: $N=1$, $N_R=0$, $N_Q=1$, RV$=1$

Sample *Sn 1* (UMUT CM16042]
Locality: Southern coast cliff of Sasage, Ôsawa Town, Chiba Prefecture [35°14.2′N, 139°52.5′E]
Horizon: Medium sand of Sanuki Formation of Kazusa Group
Age: Middle Pleistocene (ca. 0.5 Ma)
Collector: Hayami
Composition: $N=5$, $N_R=2$, $N_Q=3$, RV$=2$ (0R+2Q), LV$=3$ (2R+1Q)

Sample *Sn 2* [UMUT CM16043]
Locality: Southern coast of Sasage, Ôsawa Town, Chiba Prefecture [35°14.2′N, 139°52.5′E]
Horizon: Medium-coarse sand of Sanuki Formation of Kazusa Group
Collector: Ogose
Composition: $N=28$, $N_R=3$, $N_Q=25$, RV$=17$ (3R+14Q), LV$=11$ (0R+11Q)

Sample *Sn 3* [UMUT CM16044]
Locality: Southern coast cliff of Sasage, Ôsawa Town, Chiba Prefecture [35°14.2′N, 139°52.5′E]
Horizon: Medium sand of Sanuki Formation of Kazusa Group
Age: Middle Pleistocene (ca. 0.5 Ma)
Collector: Ozawa
Composition: $N=177$, $N_R=41$, $N_Q=136$, RV$=110$ (28R+82Q), LV$=67$ (13R+54Q)

Sample *Si* [UMUT CM16189]
Locality: Tanibori, Nishi-ôwada, Futtsu City, Chiba Prefecture
Horizon: Sand of "Sunami Formation" (Loc. 7 of Ohara, 1973)
Age: Middle Pleistocene (? ca. 0.45 Ma)
Collector: Ohta
Composition: $N=1$, $N_R=0$, $N_Q=1$, RV$=1$

Sample *Nn* [UMUT CM16190]
Locality: Tengakuin, Takaya, Fujisawa City, Kanagawa Prefecture
Horizon: Sand of "Naganuma Formation"
Age: Late Pleistocene (unknown in detail)
Collector: Ohta
Composition: $N=1$, $N_R=0$, $N_Q=1$, LV$=1$

Sample *Hg* [UMUT CM16045]
Locality: South of a pond, Kôri Dam, Kimitsu City, Chiba Prefecture (Loc. 1B by Ohara, 1973) [35°17.8′N, 139°54.7′E]
Horizon: Coarse sand of "Higashiyatsu Formation" (?=Jizôdô Formation of Shimôsa Group)
Age: Late Pleistocene?
Collector: Ozawa, Yamaguchi and Hayami
Composition: $N=55$, $N_R=9$, $N_Q=46$, RV$=47$ (8R+39Q), LV$=8$ (1R+7Q)

Sample *Ny 1* [UMUT CM16046]
Locality: Road-cut of the slope, north of Nishiyatsu, Koito Town, Chiba Prefecture [35°18.6′N, 139°59.0′E]
Horizon: Medium-coarse sand of Jizôdô Formation of Shimôsa Group
Age: Late Pleistocene (N23, ca. 0.37 Ma)
Collector: Ogose
Composition: $N=94$, $N_R=12$, $N_Q=82$, RV$=63$ (8R+55Q), LV$=31$ (4R+27Q)

Sample *Ny 2* [UMUT CM16047]
Locality: Road-cut of the slope, north of Nishiyatsu, Koito Town, Chiba Prefecture [35°18.6′N, 139°59.0′E]
Horizon: Medium-coarse sand of Jizôdô Formation of Shimôsa Group
Age: Late Pleistocene (N23, ca. 0.37 Ma)
Collector: Ohara
Composition: $N=55$, $N_R=11$, $N_Q=44$, RV$=35$ (9R+26Q), LV$=20$ (2R+18Q)

Sample *Ny 3* [UMUT CM16048]
Locality: Road-cut of the slope, north of Nishiyatsu, Koito Town, Chiba Prefecture [35°18.6′N, 139°59.0′E]
Horizon: Medium sand of Jizôdô Formation of Shimôsa Group
Age: Late Pleistocene (N23, ca. 0.37 Ma)
Collector: Yamaguchi and Hayami
Composition: $N=20$, $N_R=5$, $N_Q=15$, RV$=13$ (4R+9Q), LV$=7$ (1R+6Q)

Sample *Ny 4* [UMUT CM16049]
Locality: East of Nishiyatsu, Koito Town, Chiba Prefecture [35°18.8′N, 139°59.1′E]
Horizon: Medium sand of Jizôdô Formation of Shimôsa Group
Age: Late Pleistocene (N23, ca. 0.37 Ma)
Collector: Hayami
Composition: $N=62$, $N_R=10$, $N_Q=52$, RV$=15$ (3R+12Q), LV$=47$ (7R+40Q)

Sample *Jz* [UMUT CM16050]
Locality: Road-cut, east of a shrine, Jizôdô, Kisarazu City, Chiba Prefecture [35°22.0′N,

130

140°05.9′E]

Horizon: Coarse sand of Jizôdô Formation of Shimôsa Group

Age: Late Pleistocene (N23, ca. 0.37 Ma)

Collector: Akutsu, Yamaguchi and Hayami

Composition: $N=1731$, $N_R=218$, $N_Q=1513$, RV=819 (104R+715Q), LV=912 (114R+798Q)

Sample *Ic* [UMUT CM16079]

Locality: East of Ichinosawa, Kisarazu City, Chiba Prefecture [35°20.8′N, 140°04.2′E]

Horizon: Medium sand of Jizôdô Formation of Shimôsa Group

Age: Late Pleistocene (N23, ca. 0.37 Ma)

Collector: Ohta

Composition: $N=107$, $N_R=16$, $N_Q=91$, RV=46 (5R+41Q), LV=61 (11R+50Q)

Sample *Ab 1* [UMUT CM16051]

Locality: East of Atebi, Kisarazu City, Chiba Prefecture [35°21.4′N, 140°05.4′E]

Horizon: Medium sand of Jizôdô Formation (lower fossil bed) of Shimôsa Group

Age: Late Pleistocene (N23, ca. 0.37 Ma)

Collector: Yamaguchi and Hayami

Composition: $N=713$, $N_R=129$, $N_Q=584$, RV=326 (63R+263Q), LV=387 (66R+321Q)

Sample *Ab 2* [UMUT CM16052]

Locality: East of Atebi, Kisarazu City, Chiba Prefecture [35°41.4′N, 140°05.4′E]

Horizon: Medium sand of Jizôdô Formation (upper fossil bed) of Shimôsa Group

Age: Late Pleistocene (N23, ca. 0.37 Ma)

Collector: Yamaguchi and Hayami

Composition: $N=315$, $N_R=43$, $N_Q=272$, CV=2 (1R+1Q), RV=122 (16R+106Q), LV=191 (26R+165Q)

Sample *Nm* [UMUT CM16053]

Locality: Cliff along a small valley, southeast of Mushikubo, Ôiso Town, Kanagawa Prefecture [35°18.8′N, 139°16.7′E]

Horizon: Medium-coarse sand of Ninomiya Formation

Age: Late Pleistocene (unknown in detail)

Collector: Ozawa and Hayami

Composition: $N=24$, $N_R=6$, $N_Q=18$, RV=16 (4R+12Q), LV=8 (2R+6Q)

Sample *Sm* [UMUT CM16054]

Locality: Lower part of a cliff, Semata-no-seki, east of Ochishimoshinden, Toki Town, Chiba Prefecture [35°31.5′N, 140°13.7′E]

Horizon: Medium sand of Yabu Formation (layer E of Ohara, 1968a) of Shimôsa Group

Age: Late Pleistocene (N23, ca. 0.29 Ma)

Collector: Ohara, Akutsu and Hayami

Composition: $N=250$, $N_R=39$, $N_Q=211$, RV=126 (20R+106Q), LV=124 (19R+105Q)

Sample *Tn* [UMUT CM16055]

Locality: Iitomi, Sodegaura Town, Chiba Prefecture [35°24.9′N, 139°59.5′E]

Horizon: Medium sand of Kioroshi Formation (Toyonari Member) of Shimôsa Group

Age: Late Pleistocene (N23, ca. 0.15 Ma)

Collector: Yamaguchi

Composition: $N=2$, $N_R=0$, $N_Q=2$, RV=2 (0R+2Q)

Sample *Ys* [UMUT CM16056]

Locality: Unoki, Yoshida Village, Kagoshima Prefecture [31°44.3′N, 130°33.4′E]

Horizon: Medium-coarse sand of Yoshida Formation (Yoshida Shell Bed)

Age: Late Pleistocene (unknown in detail)

Collector: Students of Kagoshima University

Composition: $N=14$, $N_R=5$, $N_Q=9$, RV=8 (4R+4Q), LV=6 (2R+4Q)

Sample *Ms* [UMUT CM16057]

Locality: Moeshima (Niijima) Islet in Kinkô Bay, Kagoshima City, Kagoshima Prefecture [31°36.9′N, 130°43.5′E]

Horizon: Medium sand of Moeshima Shell Bed

Age: Holocene (2720±65 years B. P., according to Y. Matsushima's personal communication)

Collector: Shuto, Matsukuma, Kameyama, Okaguchi, Yamaguchi and Hayami

Composition: $N=1410$, $N_R=300$, $N_Q=1110$, RV=716 (162R+554Q), LV=694 (138R+556Q)

Other confirmed fossil specimens of *C. vesiculosus*

UMUT CM22310. "Lower Musashino Bed at Sukegawa (Tsurushihama)" [Taga Group at the south of Cape Tsurushi, Hitachi City, Ibaraki Prefecture] (described by Yokoyama, 1925)

UMUT CM21557. "Upper Musashino Bed at Shitô [probably Yabu Formation at Ochishimoshinden, Toki Town, Chiba Prefecture] (described by Yokoyama, 1922)

UMUT CM20549–20552. "Koshiba Bed at Koshiba" [Koshiba Formation at Kanazawa-shiba-machi, Kanazawa-ku, Yokohama City] (described by Yokoyama, 1911, 1920)

IGPS no. 45457. "Mandano Lignite Bed at Mandano-yama, Kimitsu County, Chiba Prefecture" (collected by Ueda)

IGPS no. 45685. "Kiyokawa Bed, Tambara, Makuda Village, Chiba Prefecture" (collected by

Ueda)

IGPS no. 46025. "Kiwada Bed at the east of Takataki River, Oikawa Village, Chiba Prefecture" (collected by Ueda)

IGPS no. 43144. "Kiwada Bed at the west of Nako Primary School, Nako, Chiba Prefecture" (collected by Ueda)

IGPS no. 24186. "Umegase Bed at Futaire, Mishima Village, Chiba Prefecture" (collected by Ueda)

IGPS no. 25162. "Umegase Bed at the east of Nishihigasa, Akimoto Village, Chiba Prefecture" (collected by Ueda)

IGPS no. 24907. "Kushiba Bed at Yamanouchi, Kosaka Village near Kamakura, Kanagawa Prefecture" (collected by Ueda)

IGPS no. 14973. "Miyata Bed at Shimomiyata, Hatsuse Village, Kanagawa Prefecture"

IGPS no register number. Hayakawa Tuff at the south of Sukumogawa, Hakone Town, Kanagawa Prefecture

IGPS no. 75331. Nekoya Formation at Nakahiramatsu, Kunô Village, Shizuoka Prefecture (collected by Nijo)

Ogasawara's private collection. Shibikawa Formation at Tayazawa, Oga City, Akita Prefecture

Ogasawara's private collection. Nobori Formation at Norobi, Muroto City, Kochi Prefecture

Kyushu University Collection (GK-L4858, 4905). Takanabe Member of Miyazaki Group at Hagenoshita, Uwaye Village, Miyazaki Prefecture and at Kizukume, Tonda Village, Miyazaki Prefecture (described by Shuto, 1973)

Recent samples

Sample Tt

  Locality: Tateyama Bay, Boso Peninsula (unknown in detail)

  Collector: unknown

  Composition: $N=23$, $N_R=8$, $N_Q=15$, RV $=11$ (3R+8Q), LV$=12$ (5R+7Q)

  Repository: Institute of Geology and Palaeontology, Tohoku University

Sample Kz [UMUT RM16058]

  Locality: Uraga Strait (mouth of Tokyo Bay), about 2 km east of Cape Kenzaki of Miura Peninsula [35°08.6′N, 139°42.0′E]

  Collector: Okutani (R/V Soyo II)

  Composition: $N=2$, $N_R=1$, $N_Q=1$, CV$=2$ (1R+1Q)

Sample Hy [UMUT RM16059]

  Locality Sagami Bay, off west coast of southern part of Miura Peninsula (several stations)

  Collector: His Majesty the Emperor of Japan

Composition (integrated): $N=352$, $N_R=159$, $N_Q=193$, CV$=38$ (20R+18Q), RV$=124$ (56R+68Q), LV$=190$ (83R+107Q)

Sample Jg

  Locality: Eastern part of Sagami Bay, about 2 km west of the western end of Jôgashima Islet (26 dredge samples at nearly the same station) [35°07.9′N, 139°35.1′E, 80–90 m]

  Collector: Hayami et al (R/V Rinkai)

  Composition (integrated): $N=10734$, $N_R=4547$, $N_Q=6187$, CV$=228$ (113R+115Q), RV$=5370$ (2235R+3135Q), LV$=5136$ (2199R+2937Q)

[Subsample Jg (*1*)] [UMUT RM16060] $N=524$, $N_R=215$, $N_Q=309$, CV$=10$ (8R+2Q), RV$=268$ (103R+165Q), LV$=246$ (104R+142Q)

[Subsample Jg (*2*)] [UMUT RM16061] $N=729$, $N_R=305$, $N_Q=424$, CV$=21$ (10R+11Q), RV$=375$ (158R+217Q), LV$=333$ (137R+196Q)

[Subsample Jg (*3*)] [UMUT RM16062] $N=807$, $N_R=337$, $N_Q=470$, CV$=11$ (8R+3Q), RV$=408$ (166R+242Q), LV$=388$ (163R+225Q)

[Subsample Jg (*4*)] [UMUT RM16063] $N=669$, $N_R=280$, $N_Q=389$, CV$=18$ (10R+8Q), RV$=345$ (130R+215Q), LV$=306$ (140R+166Q)

[Subsample Jg (*5*)] [UMUT RM16064] $N=358$, $N_R=170$, $N_Q=188$, CV$=5$ (3R+2Q), RV$=178$ (92R+86Q), LV$=175$ (75R+100Q)

[Subsample Jg (*6*)] [UMUT RM16065] $N=564$, $N_R=224$, $N_Q=340$, CV$=4$ (1R+3Q), RV$=227$ (114R+163Q), LV$=283$ (109R+174Q)

[Subsample Jg (*7*)] [UMUT RM16066] $N=1078$, $N_R=437$, $N_Q=641$, CV$=29$ (11R+18Q), RV$=526$ (211R+315Q), LV$=523$ (215R+308Q)

[Subsample Jg (*8*)] [UMUT RM16067] $N=1838$, $N_R=792$, $N_Q=1046$, CV$=29$ (12R+17Q), RV 881 (364R+517Q), LV$=928$ (416R+512Q)

[Subsample Jg (*9*)] [UMUT RM16068] $N=1606$, $N_R=730$, $N_Q=876$, CV$=33$ (16R+17Q), RV $=797$ (347R+450Q), LV$=776$ (367R+409Q)

[Subsample Jg (*10*)] [UMUT RM16069] $N=621$, $N_R=242$, $N_Q=379$, CV$=3$ (1R+2Q), RV$=329$ (127R+202Q), LV$=289$ (114R+175Q)

[Subsample Jg (*11*)] [UMUT RM16070] $N=567$, $N_R=245$, $N_Q=322$, CV$=1$ (0R+1Q), RV$=288$ (125R+163Q), LV$=278$ (120R+158Q)

[Subsample Jg (*12*)] [UMUT RM16071] $N=87$, $N_R=32$, $N_Q=55$, CV$=2$ (2R+0Q), RV$=45$ (16R+29Q), LV$=40$ (14R+26Q)

[Subsample Jg (*13*)] [UMUT RM16072] $N=510$, $N_R=223$, $N_Q=287$, CV$=2$ (2R+0Q), RV$=275$ (133R+142Q), LV$=255$ (88R+145Q)

[Subsample Jg (*14*)] [UMUT RM16073] $N=322$,

132

$N_R=118$, $N_Q=204$, CV=1 (1R+0Q), RV= 170 (63R+107Q), LV=151 (54R+97Q)

[Subsample *Jg (15)*] [UMUT RM16074] $N=396$, $N_R=169$, $N_Q=227$, CV=1 (0R+1Q), RV= 208 (86R+122Q), LV=187 (83R+104Q)

[Subsample *Jg (16)*] [UMUT RM16172] $N=4$, $N_R=2$, $N_Q=2$, CV=4 (2R+2Q) (Dead shells numerous but not counted)

[Subsample *Jg (17)*] [UMUT RM16173] $N=6$, $N_R=2$, $N_Q=4$, CV=6 (2R+4Q) (Dead shells numerous but not counted)

[Subsample *Jg (18)*] [UMUT RM16174] $N=5$, $N_R=5$, $N_Q=0$, CV=5 (5R+0Q) (Dead shells numerous but not counted)

[Subsample *Jg (19)*] [UMUT RM16175] $N=11$, $N_R=4$, $N_Q=7$, CV=11 (4R+7Q) (Dead shells numerous but not counted)

[Subsample *Jg (20)*] [UMUT RM16176] $N=19$, $N_R=8$, $N_Q=11$, CV=19 (8R+11Q) (Dead shells numerous but not counted)

[Subsample *Jg (21)*] [UMUT RM16177] $N=13$, $N_R=7$, $N_Q=2$, CV=13 (7R+6Q) (Dead shells numerous but not counted)

[Subsample *Jg (22)*] [UMUT RM16178] $N=5$, $N_R=3$, $N_Q=2$, CV=5 (3R+2Q) (Dead shells numerous but not counted)

[Subsample *Jg (23)*] [UMUT RM16179] $N=8$, $N_R=1$, $N_Q=7$, CV=8 (1R+7Q) (Dead shells numerous but not counted)

[Subsample *Jg (24)*] [UMUT RM16191] $N=10$, $N_R=4$, $N_Q=6$, CV=10 (4R+6Q) (Dead shells numerous but not counted)

[Subsample *Jg (25)*] [UMUT RM16192] $N=10$, $N_R=6$, $N_Q=4$, CV=10 (6R+4Q) (Dead shells numerous but not counted)

[Subsample *Jg (26)*] [UMUT RM16193] $N=6$, $N_R=2$, $N_Q=4$, CV=6 (2R+4Q) (Dead shells numerous but not counted)

Sample *Ma* [UMUT RM16075]

Locality: Eastern part of Sagami Bay, 5 km west of the western end of Jôgashima Islet [35°08.5′N, 139°33.7′E, 105 m]

Collector: Hayami et al. (R/V Rinkai)

Composition: $N=70$, $N_R=26$, $N_Q=44$, CV= 1 (1R+0Q), RV=43 (14R+29Q), LV=26 (11R+15Q)

Sample *Am* [UMUT RM16076]

Locality: Eastern part of Sagami Bay, Minami-amadaiba about 7 km west of Cape Arasaki [35°11.6′N, 139°32.0′E, 110 m]

Collector: Hayami et al. (R/V Rinkai)

Composition: $N=223$, $N_R=87$, $N_Q=136$, RV =125 (50R+75Q), LV=98 (37R+61Q)

Sample *Aj*

Locality: Western part of Sagami Bay, off the coast between Ajiro and Usami of Izu Peninsula (2 stations)

Collector: Okutani (R/V Soyo II)

Composition (integrated): $N=3$, $N_R=1$, $N_Q=2$, CV=3 (1R+2Q)

[Subsample *Aj (1)*] [UMUT RM16077] (T10 by Okutani, 35°00.5′N, 139°08.1′E, 105 m) $N=2$, $N_R=0$, $N_Q=2$, CV=2 (0R+2Q)

[Subsample *Aj (2)*] [UMUT RM16078] (T18 by Okutani, 35°02.5′N, 139°08.0′E, 90 m) $N=1$, $N_R=1$, $N_Q=0$, CV=1 (1R+0Q)

Sample *Os*

Locality: Sagami Bay, off Okada Harbor, Oshima (KT73–5 St. T (2nd), 34°48′20″N, 139°23′67″E–34°48′42″N, 139°23′02″E, 109–116 m)

Collector: Ohta (R/V Tansei)

Composition: $N=2$, $N_R=2$, $N_Q=0$, CV=2 (2R+0Q)

Repository: Ocean Research Institute, University of Tokyo

Sample *Zs*

Locality: Zenisu, a sea bank southwest of Kôzu Island (9 stations)

Collector: Okutani (R/V Sôyô II)

Composition (integrated): $N=101$, $N_R=40$, $N_Q=61$, CV=9 (5R+4Q), RV=59 (24R+35Q), LV=33 (11R+22Q)

[Subsample *Zs (1)*] [UMUT RM16080] (St. 32 by Okutani, 1972, 34°00.8′N, 138°50.8′E, 113 m) $N=14$, $N_R=3$, $N_Q=11$, CV=1 (0R+1Q), RV =9 (3R+6Q), LV=4 (0R+4Q)

[Subsample *Zs (2)*] [UMUT RM16081] (D21 by Okutani, 1972, 34°02.1′N, 138°51.5′E, 125 m) $N=2$, $N_R=1$, $N_Q=1$, RV=1 (1R+0Q), LV= 1 (0R+1Q)

[Subsample *Zs (3)*] [UMUT RM16082] (D22 by Okutani, 1972, 34°00.8′N, 138°50.8′E, 110 m) $N=8$, $N_R=6$, $N_Q=2$, CV=2 (2R+0Q), RV=4 (2R+2Q), LV=2 (2R+0Q)

[Subsample *Zs (4)*] [UMUT RM16083] (D23 by Okutani, 1972, 34°00.1′N, 138°51.0′E, 90 m) $N=20$, $N_R=8$, $N_Q=12$, CV=4 (2R+2Q), RV =8 (4R+4Q), LV=8 (2R+6Q)

[Subsample *Zs (5)*] [UMUT RM16084] (D24 by Okutani, 1972, 33°59.1′N, 138°50.9′E, 80 m) $N =7$, $N_R=3$, $N_Q=4$, RV=1 (1R+0Q), LV=6 (2R+4Q)

[Subsample *Zs (6)*] [UMUT RM16085] (D47 by Okutani, 1972, 33°53.9′N, 138°51.0′E, 150 m) $N=10$, $N_R=4$, $N_Q=6$, RV=8 (3R+5Q), LV=

2 (1R+1Q)

[Subsample *Zs* (*7*)] [UMUT RM16086] (D49 by Okutani, 1972, 33°58.7′N, 138°51.5′E, 100 m) $N=7$, $N_R=1$, $N_Q=6$, RV=6 (1R+5Q), LV=1 (0R+1Q)

[Subsample *Zs* (*8*)] [UMUT RM16087] (D50 by Okutani, 1972, 34°00.6′N, 138°50.2′E, 106 m) $N=10$, $N_R=3$, $N_Q=7$, CV=2 (1R+1Q), RV=4 (1R+3Q), LV=4 (1R+3Q)

[Subsample *Zs* (*9*)] [UMUT RM16088] (unknown in detail) $N=23$, $N_R=11$, $N_Q=12$, RV=18 (8R+10Q), LV=5 (3R+2Q)

Sample *Hs*

Locality: Hyôtanse, a sea bank west of Niijima Island (8 stations)

Collector: Okutani (R/V Sôyô II)

Composition (integrated): $N=45$, $N_R=20$, $N_Q=25$, CV=1 (0R+1Q), RV=26 (10R+16Q), LV=18 (10R+8Q)

[Subsample *Hs* (*1*)] [UMUT RM16089] (St. 16 by Okutani, 1972, 34°22.0′N, 139°04.4′E, 125 m) $N=2$, $N_R=1$, $N_Q=1$, RV=1 (1R+0Q), LV=1 (0R+1Q)

[Subsample *Hs* (*2*)] [UMUT RM16090] (St. 25 (1) by Okutani, 1972, 34°20.2′N, 139°02.8′E, 120–125 m) $N=1$, $N_R=0$, $N_Q=1$, CV=1 (0R+1Q)

[Subsample *Hs* (*3*)] [UMUT RM16091] (D15 by Okutani, 1972, 34°21.1′N, 139°03.8′E, 170 m) $N=4$, $N_R=4$, $N_Q=0$, RV=1 (1R+0Q), LV=3 (3R+0Q)

[Subsample *Hs* (*4*)] [UMUT RM16092] (D16 by Okutani, 1972, 34°22.2′N, 139°05.0′E, 140 m) $N=8$, $N_R=3$, $N_Q=5$, RV=5 (2R+3Q), LV=3 (1R+2Q)

[Subsample *Hs* (*5*)] [UMUT RM16093] (D28 by Okutani, 1972, 34°20.9′N, 139°03.6′E, 115–120 m) $N=9$, $N_R=3$, $N_Q=6$, RV=6 (1R+5Q), LV=3 (2R+1Q)

[Subsample *Hs* (*6*)] [UMUT RM16094] (D52 by Okutani, 1972, 34°20.9′N, 139°03.0′E, 135 m) $N=8$, $N_R=5$, $N_Q=3$, RV=6 (3R+3Q), LV=2 (2R+5Q)

[Subsample *Hs* (*7*)] [UMUT RM16095] (D53 by Okutani, 1972, 34°22.1′N, 139°05.4′E, 117 m) $N=11$, $N_R=4$, $N_Q=7$, RV=6 (2R+4Q), LV=5 (2 R+3Q)

[Subsample *Hs* (*8*)] [UMUT RM16096] (D55 by Okutani, 1972, 34°21.0′N, 139°05.7′E, 220 m) $N=2$, $N_R=0$, $N_Q=2$, RV=1 (0R+1Q), LV=1 (0R+1Q)

Sample *Ts*

Locality: Takase, a sea bank northwest of Niijima Island (11 stations)

Collector: Okutani (R/V Sôyô II) and Hayami et al. (R/V Tansei)

Composition (integrated): $N=241$, $N_R=107$, $N_Q=134$, CV=3 (1R+2Q), RV=128 (60R+68Q), LV=110 (46R+64Q)

[Subsample *Ts* (*1*)] [UMUT RM16097] (D17 by Okutani, 1972, 34°26.1′N, 139°11.9′E, 80–110 m), $N=1$, $N_R=0$, $N_Q=1$, RV=1 (0R+1Q)

[Subsample *Ts* (*2*)] [UMUT RM16098] (D19 by Okutani, 1972, 34°27.5′N, 139°12.4′E, 100 m) $N=9$, $N_R=7$, $N_Q=2$, CV=1 (0R+1Q), RV=6 (5R+1Q), LV=2 (2R+0Q)

[Subsample *Ts* (*3*)] [UMUT RM16099] (D31 by Okutani, 1972, 34°27.5′N, 139°12.5′E, 120–200 m) $N=2$, $N_R=0$, $N_Q=2$, CV=1 (0R+1Q), RV=1 (0R+1Q)

[Subsample *Ts* (*4*)] [UMUT RM16100] (D32 by Okutani, 1972, 34°28.3′N, 139°11.4′E, 110 m) $N=4$, $N_R=1$, $N_Q=3$, RV=3 (1R+2Q), LV=1 (0R+1Q)

[Subsample *Ts* (*5*)] [UMUT RM16101] (D34 by Okutani, 1972, 34°26.5′N, 139°10.3′E, 70 m) $N=3$, $N_R=2$, $N_Q=1$, CV=1 (1R+0Q), RV=1 (0R+1Q), LV=1 (1R+0Q)

[Subsample *Ts* (*6*)] [UMUT RM16102] (D35 by Okutani, 1972, 34°26.8′N, 139°12.2′E, 95–105 m) $N=5$, $N_R=3$, $N_Q=2$, RV=3 (1R+2Q), LV=2 (2R+0Q)

[Subsample *Ts* (*7*)] [UMUT RM16103] (D56 by Okutani, 1972, 34°27.7′N, 139°10.0′E, 180–200 m) $N=6$, $N_R=2$, $N_Q=4$, RV=3 (1R+2Q), LV=3 (1R+2Q)

[Subsample *Ts* (*8*)] [UMUT RM16104] (D58 by Okutani, 1972, 34°26.7′N, 139°11.3′E, 95 m) $N=65$, $N_R=25$, $N_Q=40$, RV=39 (15R+24Q), LV=26 (10R+16Q)

[Subsample *Ts* (*9*)] [UMUT RM16105] (D59 by Okutani, 1972, 34°28.4′N, 139°11.2′E, 95 m) $N=22$, $N_R=9$, $N_Q=13$, RV=6 (4R+2Q), LV=16 (5R+11Q)

[Subsample *Ts* (*10*)] [UMUT RM16106] (D60 by Okutani, 1972, 34°28.4′N, 139°12.9′E, 170–210 m) $N=8$, $N_R=3$, $N_Q=5$, RV=5 (2R+3Q), LV=3 (1R+2Q)

[Subsample *Ts* (*11*)] [UMUT RM16107] (KT78-2–Ta 1, 34°27.6′N, 139°12.4′E, 100 m) $N=116$, $N_R=55$, $N_Q=61$, RV=60 (31R+29Q), LV=56 (24R+32Q)

Sample *Hk* [UMUT RM16108]

Locality: Eastern part of Suruga Bay, off Cape Hakachi (KT81-5-T–110, 34°55.0′N, 138°46.1′E, 110 m)

Collector: Kitazato et al. (R/V Tansei)

134

Composition: $N=4$, $N_R=1$, $N_Q=3$, CV=4 (1R+3Q)

Sample *Su*

Locality: Senoumi, a sea bank in western part of Suruga Bay (29 stations)

Collector: Sanyo Suiro Co. and Imajima et al. (R/V Tansei)

Composition (integrated): $N=1650$, $N_R=660$, $N_Q=990$, RV=773 (320R+453Q), LV=877 (340R+537Q)

[Subsample *Su* (*1*)] [UMUT RM16109] (St. 2 by Sanyo Suiro, 34°45′15″N, 138°31′48″E, 137 m) $N=10$, $N_R=2$, $N_Q=8$, RV=6 (1R+5Q), LV= 4 (1R+3Q)

[Subsample *Su* (*2*)] [UMUT RM16110] (St. 3 by Sanyo Suiro, 34°45′04″N, 138°31′38″E, 107 m) $N=18$, $N_R=6$, $N_Q=12$, RV=10 (1R+9Q), LV=8 (5R+3Q)

[Subsample *Su* (*3*)] [UMUT RM16111] (St. 4 by Sanyo Suiro, 34°45′04″N, 138°31′38″E, 69 m) $N=72$, $N_R=29$, $N_Q=43$, RV=31 (15R+16Q), LV=41 (14R+27Q)

[Subsample *Su* (*4*)] [UMUT RM16112] (St. 7 by Sanyo Suiro, 24°45′03″N, 138°30′25″E, 74 m) $N=29$, $N_R=9$, $N_Q=20$, RV=14 (6R+8Q), LV =15 (3R+12Q)

[Subsample *Su* (*5*)] [UMUT RM16113] (St. 8 by Sanyo Suiro, 34°44′50″N, 138°30′31″E, 60 m) $N=7$, $N_R=2$, $N_Q=5$, RV=4 (2R+2Q), LV=3 (0R+3Q)

[Subsample *Su* (*6*)] [UMUT RM16114] (St. 9 by Sanyo Suiro, 34°43′45″N, 138°30′31″E, 61 m) $N=13$, $N_R=5$, $N_Q=8$, RV=7 (4R+3Q), LV= 6 (1R+5Q)

[Subsample *Su* (*7*)] [UMUT RM16115] (St. 11 by Sanyo Suiro, 34°44′46″N, 138°31′24″E, 54 m) $N=10$, $N_R=4$, $N_Q=6$, RV=4 (1R+3Q), LV =6 (3R+3Q)

[Subsample *Su* (*8*)] [UMUT RM16116] (St. 12 by Sanyo Suiro, 34°44′46″N, 138°31′49″E, 61 m) $N=9$, $N_R=4$, $N_Q=5$, RV=4 (2R+2Q), LV=5 (2R+3Q)

[Subsample *Su*(*9*)][UMUT RM16117] (St. 13 by Sanyo Suiro, 34°44′22″N, 138°32′16″E, 84 m) $N=157$, $N_R=67$, $N_Q=90$, RV=85 (42R+43Q), LV=72 (25R+47Q)

[Subsample *Su* (*10*)][UMUT RM16118] (St. 15 by Sanyo Suiro, 34°44′28″N, 138°31′19″E, 56 m) $N=95$, $N_R=41$, $N_Q=54$, RV=44 (22R+22Q), LV=51 (19R+32Q)

[Subsample *Su* (*11*)] [UMUT RM16119] (St. 16 by Sanyo Suiro, 34°44′32″N, 138°30′53″E, 48 m) $N=8$, $N_R=3$, $N_Q=5$, RV=1 (1R+0Q), LV=7 (2R+5Q)

[Subsample *Su* (*12*)] [UMUT RM16120] (St. 17 by Sanyo Suiro, 34°44′26″N, 138°30′24″E, 53 m) $N=69$, $N_R=29$, $N_Q=40$, RV=27 (7R+20Q), LV=42 (22R+20Q)

[Subsample *Su* (*13*)] [UMUT RM16121] (St. 18 by Sanyo Suiro, 34°44′22″N, 138°29′57″E, 63 m) $N=109$, $N_R=47$, $N_Q=62$, RV=41 (18R+23Q), LV=68 (29R+39Q)

[Subsample *Su* (*14*)] [UMUT RM16122] (St. 19 by Sanyo Suiro, 34°43′55″N, 138°29′27″E, 64 m) $N=11$, $N_R=4$, $N_Q=7$, RV=7 (3R+4Q), LV =4 (1R+3Q)

[Subsample *Su* (*15*)] [UMUT RM16123] (St. 20 by Sanyo Suiro, 34°43′58″N, 138°30′30″E, 57 m) $N=9$, $N_R=5$, $N_Q=4$, RV=3 (2R+1Q), LV= 6 (3R+3Q)

[Subsample *Su* (*16*)] [UMUT RM16124] (St. 21 by Sanyo Suiro, 34°43′57″N, 138°30′47″E, 57 m) $N=31$, $N_R=13$, $N_Q=18$, RV=15 (6R+9Q), LV=16 (7R+9Q)

[Subsample *Su* (*17*)] [UMUT RM16125] (St. 22 by Sanyo Suiro, 34°43′59″N, 138°31′24″E, 59 m) $N=12$, $N_R=4$, $N_Q=8$, RV=6 (2R+4Q), LV =6 (2R+4Q)

[Subsample *Su* (*18*)] [UMUT RM16126] (St. 23 by Sanyo Suiro, 34°44′14″N, 138°31′27″E, 64 m) $N=3$, $N_R=0$, $N_Q=3$, RV=1 (0R+1Q), LV=2 (0R+2Q)

[Subsample *Su* (*19*)] [UMUT RM16127] (St. 24 by Sanyo Suiro, 34°43′59″N, 138°32′09″E, 63 m) $N=4$, $N_R=0$, $N_Q=4$, RV=2 (0R+2Q), LV=2 (0R+2Q)

[Subsample *Su* (*20*)] [UMUT RM16128] (St. 25 by Sanyo Suiro, 34°43′49″N, 138°30′07″E, 70 m) $N=110$, $N_R=42$, $N_Q=68$, RV=47 (17R+30Q), LV=63 (25R+38Q)

[Subsample *Su* (*21*)] [UMUT RM16129] (St. 26 by Sanyo Suiro, 34°43′35″N, 138°30′54″E, 64 m) $N=100$, $N_R=51$, $N_Q=49$, RV=45 (23R 22Q), LV=55 (28R+27Q)

[Subsample *Su* (*22*)] [UMUT RM16130] (St. 27 by Sanyo Suiro, 34°43′40″N, 138°31′20″E, 74 m) $N=110$, $N_R=41$, $N_Q=69$, RV=59 (18R+41Q), LV=51 (23R+28Q)

[Subsample *Su* (*23*)] [UMUT RM16131] (St. 28 by Sanyo Suiro, 34°43′44″N, 138°31′49″E, 71 m) $N=54$, $N_R=20$, $N_Q=34$, RV=22 (8R+14Q), LV=32 (12R+20Q)

[Subsample *Su* (*24*)] [UMUT RM16132] (St. 29 by Sanyo Suiro, 34°43′27″N, 138°32′11″E, 95 m) $N=24$, $N_R=11$, $N_Q=13$, RV=12 (5R+7Q), LV=12 (6R+6Q)

[Subsample *Su* (*25*)] [UMUT RM16133] (St. 30 by Sanyo Suiro, 34°43′21″N, 138°32′17″E, 106 m) $N=27$, $N_R=7$, $N_Q=20$, RV=11 (2R+9Q), LV=16 (5R+11Q)

[Subsample *Su* (*26*)] [UMUT RM16134] (St. 31 by Sanyo Suiro, 34°43′16″N, 138°31′08″E, 87 m) $N=20$, $N_R=10$, $N_Q=10$, RV=9 (3R+6Q), LV=11 (7R+4Q)

[Subsample *Su* (*27*)] [UMUT RM16135] (St. 32 by Sanyo Suiro, 34°43′32″N, 138°29′49″E, 105 m) $N=22$, $N_R=7$, $N_Q=15$, RV=13 (6R+7Q), LV=9 (1R+8Q)

[Subsample *Su* (*28*)] [UMUT RM16136] (St. 33 by Sanyo Suiro, 34°43′19″N, 138°29′32″E, 75 m) $N=14$, $N_R=6$, $N_Q=8$, RV=7 (2R+5Q), LV=7 (4R+3Q)

[Subsample *Su* (*29*)] [UMUT RM16137] (KT78-2-Z6, 34°43.5′–34°43.8′N, 138°30.6′E, 70–71 m) $N=493$, $N_R=188$, $N_Q=305$, RV=236 (101R+135Q), LV=257 (90R+167Q)

Sample *Is* [UMUT RM16138]
Locality: Mouth of Ise Bay off Cape Anori of Shima Peninsula (ca. 100 m)
Collector: Hayashi (through fishery at Isshiki Town in Aichi Prefecture)
Composition (integrated): $N=312$, $N_R=143$, $N_Q=169$, CV=312 (143R+169Q)

Sample *Km*
Locality: Shelf off Kushimoto and Shimoura, southern part of Kii Peninsula (11 stations)
Collector: Tsuchida
Composition (integrated): $N=131$, $N_R=44$, $N_Q=87$, CV=4 (1R+3Q), RV=61 (20R+41Q), LV=66 (23R+43Q)

[Subsample *Km* (*1*)] [UMUT RM16139] (off Kushimoto, 110–115 m) $N=89$, $N_R=31$, $N_Q=58$, RV=41 (13R+28Q), LV=48 (18R+30Q)

[Subsample *Km* (*2*)] [UMUT RM16140] (off Kushimoto, 80 m) $N=26$, $N_R=9$, $N_Q=17$, RV=17 (5R+12Q), LV=9 (4R+5Q)

[Subsample *Km* (*3*)] [UMUT RM16141] (off Kushimoto, 80 m) $N=4$, $N_R=1$, $N_Q=3$, LV=4 (1R+3Q)

[Subsample *Km* (*4*)] [UMUT RM16142] (off Kushimoto, 100 m) $N=2$, $N_R=1$, $N_Q=1$, CV=2 (1R+1Q)

[Subsample *Km* (*5*)] [UMUT RM16143] (off Kushimoto, 120 m) $N=2$, $N_R=0$, $N_Q=2$, RV=1 (0R+1Q), LV=1 (0R+1Q)

[Subsample *Km* (*6*)] [UMUT RM16144] (off Shimoura, 60 m) $N=2$, $N_R=2$, $N_Q=0$, RV=2 (2R+0Q)

[Subsample *Km* (*7*)] [UMUT RM16145] (off Kamiura, 100 m) $N=1$, $N_R=0$, $N_Q=1$, LV=1 (0R+1Q)

[Subsample *Km* (*8*)] [UMUT RM16146] (off Shimoura, 150–160 m) $N=1$, $N_R=0$, $N_Q=1$, LV=1 (0R+1Q)

[Subsample *Km* (*9*)] [UMUT RM16147] (off Kushimoto, 100 m) $N=2$, $N_R=0$, $N_Q=2$, CV=2 (0R+2Q)

[Subsample *Km* (*10*)] [UMUT RM16148] (off Shimoura, 100 m) $N=1$, $N_R=0$, $N_Q=1$, LV=1 (0R+1Q)

[Subsample *Km* (*11*)] [UMUT RM16149] (off Kushimoto, 110 m) $N=1$, $N_R=0$, $N_Q=1$, LV=1 (0R+1Q)

Sample *Sr* [UMUT RM16186]
Locality: About 6 km SSW of Setozaki, Shirahama, Kii Peninsula (33°36.0′N, 135°18′E, 140–160 m)
Collector: Yamaguchi, Oji and Kamiya
Composition: $N=14$, $N_R=2$, $N_Q=12$, CV=4 (0R+4Q), RV=1 (0R+1Q), LV=9 (2R+7Q)

Sample *Mb* [UMUT RM16150]
Locality: Mouth of Tanabe Bay (unknown in detail)
Collector: Yamaguchi (through fishery at Sakai of Minabe Town, Wakayama Prefecture)
Composition: $N=1$, $N_R=0$, $N_Q=1$, CV=1 (0R+1Q)

Sample *Bt* [UMUT RM16151]
Locality: Mouth of Kii Strait, off Bentenjima Islet (33°47.4′N, 134°50.0′E, 80 m)
Collector: Yamaguchi (R/V Shumpu)
Composition: $N=73$, $N_R=22$, $N_Q=51$, RV=44 (13R+31Q), LV=29 (9R+20Q)

Sample *Ks* [UMUT RM16152]
Locality: Mouth of Kii Strait (unknown in detail)
Collector: Hayami (through fishery at Tatsugahama of Arida City, Wakayama Prefecture)
Composition: $N=7$, $N_R=3$, $N_Q=4$, CV=4 (2R+2Q), RV=3 (1R+2Q)

Sample *Az* [UMUT RM16153]
Locality: Shelf off Cape Ashizuri, western Shikoku (unknown in detail)
Collector: Kurohara (through fishery at Shimizu City in Kochi Prefecture)
Composition: $N=4$, $N_R=0$, $N_Q=4$, CV=2 (0R+2Q), RV=1 (0R+1Q), LV=1 (0R+1Q)

Sample *Yk* [UMUT RM16154]
Locality: Coast of Cape Nagasaki-bana, Yamakawa Town, Kagoshima Prefecture [31° 09.2′N, 130°35.3′E]
Collector: Hayami

Composition: $N=2$, $N_R=1$, $N_Q=1$, RV$=2$ (1R$+$1Q)

Sample *Ec*

Locality: East China Sea, west of Tokara Islands (2 stations)

Collector: Otsuka (R/V Kagoshima)

Composition (integrated): $N=824$, $N_R=246$, $N_Q=578$, RV$=417$ (124R$+$293Q), LV$=407$ (122R$+$285Q)

[Subsample *Ec* (*1*)] [UMUT RM16155] (29°09.9′N, 126°28.1′E, ca. 100 m) $N=388$, $N_R=121$, $N_Q=267$, RV$=206$ (67R$+$139Q), LV$=182$ (54R$+$128Q)

[Subsample *Ec* (*2*)] [UMUT RM16156] (29°10.1′N, 126°28.9′E, ca. 100 m) $N=436$, $N_R=125$, $N_Q=311$, RV$=211$ (57R$+$154Q), LV$=225$ (68R$+$157Q)

Sample *My*

Locality: Off Miyakojima Island, East China Sea (unknown in detail)

Collector: unknown

Composition: $N=17$, $N_R=3$, $N_Q=14$, RV$=5$ (0R$+$5Q), LV$=12$ (3R$+$9Q)

Repository: Department of Zoology, National Science Museum, Tokyo (NSMT-49433)

Sample *Sk* [UMUT RM16182]

Locality: northeast of Senkaku Islets, East China Sea (KH68–2–T–12, 26°55′N, 125°01′E, 120 m)

Collector: Tsuchida (R/V Hakuho)

Composition: $N=89$, $N_R=8$, $N_Q=81$, CV$=7$ (1R$+$6Q), RV$=43$ (1R$+$42Q), LV$=39$ (6R$+$33Q)

Sample *Kj*

Locality East China Sea, west of western coast of Kyushu (9 stations)

Collector: Oyama (R/V Tokai II)

Composition (integrated): $N=332$, $N_R=87$, $N_Q=245$, RV$=173$ (42R$+$131Q), LV$=159$ (45R$+$114Q)

[Subsample *Kj* (*1*)] [UMUT RM16157] (Loc. 141, 31°44.2′N, 129°42.5′E, 150 m) $N=3$, $N_R=0$, $N_Q=3$, RV$=3$ (1R$+$2Q)

[Subsample *Kj* (*2*)] [UMUT RM16158] (Loc. 296, 32°49.5′N, 128°07.5′E, 100 m) $N=11$, $N_R=5$, $N_Q=6$, RV$=5$ (2R$+$3Q), LV$=6$ (3R$+$3Q)

[Subsample *Kj* (*3*)] [UMUT RM16159] (Loc. 306, 32°29.0′N, 129°42.5′E, 77 m) $N=1$, $N_R=0$, $N_Q=1$, RV$=1$ (0R$+$1Q)

[Subsample *Kj* (*4*)] [UMUT RM16160] (Loc. 308, 32°20.6′N, 129°31.9′E, 93 m) $N=20$, $N_R=5$, $N_Q=15$, RV$=10$ (2R$+$8Q), LV$=10$ (3R$+$7Q)

[Subsample *Kj* (*5*)] [UMUT RM16161] (Loc. 331, 32°20.1′N, 129°27.4′E, 86 m) $N=10$, $N_R=3$, $N_Q=7$, RV$=5$ (1R$+$4Q), LV$=5$ (2R$+$3Q)

[Subsample *Kj* (*6*)] [UMUT RM16162] (Loc. 334, 32°47.4′N, 129°16.2′E, 103 m) $N=54$, $N_R=17$, $N_Q=38$, RV$=22$ (5R$+$17Q), LV$=32$ (12R$+$20Q)

[Subsample *Kj* (*7*)] [UMUT RM16163] (Loc. 339, 33°08.7′N, 129°14.7′E, 69 m) $N=113$, $N_R=24$, $N_Q=89$, RV$=61$ (13R$+$48Q), LV$=52$ (11R$+$41Q)

[Subsample *Kj* (*8*)] [UMUT RM16164] (Loc. 340, 33°13.7′N, 129°16.6′E, 85 m) $N=98$, $N_R=25$, $N_Q=73$, RV$=54$ (15R$+$39Q), LV$=44$ (10R$+$34Q)

[Subsample *Kj* (*9*)] [UMUT RM16165] (Loc. 341, 32°47.7′N, 129°34.0′E, 80 m) $N=22$, $N_R=7$, $N_Q=15$, RV$=12$ (3R$+$9Q), LV$=10$ (4R$+$6Q)

Sample *Ng* [UMUT RM16166]

Locality: East China Sea, off Nagasaki (unknown in detail)

Collector: Yamamoto (through fishery at Mogi in Nagasaki City)

Composition: $N=10$, $N_R=6$, $N_Q=4$, CV$=10$ (6R$+$4Q)

Sample *Ta*

Locality: Tsushima Strait, south and east of Tsushima Islands (8 stations)

Collector: Habe and Imajima (R/V Genkai)

Composition (integrated): $N=1391$, $N_R=350$, $N_Q=1041$, CV$=1$ (1R$+$0Q), RV$=688$ (176R$+$512Q), LV$=702$ (173R$+$529Q)

Repository: Department of Zoology, National Science Museum, Tokyo (not yet registered)

[Subsample *Ta* (*1*)] (St. 10 by Imajima, 1970, 33°56.8′N, 129°11.9′E, 105 m) $N=301$, $N_R=68$, $N_Q=233$, RV$=132$ (31R$+$101Q), LV$=169$ (37R$+$132Q)

[Subsample *Ta* (*2*)] (St. 27 by Imajima, 1970, 34°44.0′N, 129°47.1′E, 105 m) $N=318$, $N_R=74$, $N_Q=244$, RV$=164$ (36R$+$128Q), LV$=154$ (38R$+$116Q)

[Subsample *Ta* (*3*)] (St. 11 by Imajima, 1970, 34°02.8′N, 129°04.5′E, 115 m) $N=187$, $N_R=47$, $N_Q=140$, RV$=99$ (25R$+$74Q), LV$=88$ (22R$+$66Q)

[Subsample *Ta* (*4*)] (St. 8 by Imajima, 1970, 33°49.7′N, 129°28.9′E, 100 m) $N=272$, $N_R=87$, $N_Q=185$, RV$=140$ (44R$+$96Q), LV$=132$ (43R$+$89Q)

[Subsample *Ta* (*5*)] (St. 35 by Imajima, 1970, 34°35.6′N, 130°00.0′E, 115 m) $N=76$, $N_R=20$, $N_Q=56$, CV$=1$ (1R$+$0Q), RV$=32$ (11R$+$21Q), LV$=43$ (8R$+$35Q)

[Subsample *Ta* (*6*)] (St. 31 by Imajima, 1970, 34°31.9′N, 129°36.6′E, 96 m) $N=97$, $N_R=22$, $N_Q=75$, RV=48 (11R+37Q), LV=49 (11R+38Q)

[Subsample *Ta* (*7*)] (St. 28 by Imajima, 1970, 34°45.8′N, 129°56.8′E, 120 m) $N=102$, $N_R=24$, $N_Q=78$, RV=55 (11R+44Q), LV=47 (13R+34Q)

[Subsample *Ta* (*8*)] (St. 36 by Imajima, 1970, 34°15.9′N, 130°05.8′E, 95 m) $N=38$, $N_R=8$, $N_Q=30$, RV=18 (7R+11Q), LV=20 (1R+19Q)

Sample *Hm* [UMUT RM16167]
  Locality: Japan Sea, northwest of Hamada (D220, 35°21.4′N, 131°15.4′E, 106–107 m)
  Collector: Ogasawara et al. (R/V Hakurei)
  Composition: $N=115$, $N_R=25$, $N_Q=90$, RV=50 (13R+37Q), LV=65 (12R+53Q)

Sample *Im* [UMUT RM16168]
  Locality: Japan Sea, off Izumo Peninsula (D306, 35°42.7′N, 132°47.5′E, 120–125 m)
  Collector: Ogasawara et al. (R/V Hakurei)
  Composition: $N=63$, $N_R=15$, $N_Q=48$, RV=38 (8R+30Q), LV=25 (7R+18Q)

Sample *Nt* [UMUT RM16169]
  Locality: Japan Sea, off Cape Suzu, Noto Peninsula (KT79–5–16D, 37°23.2′N, 137°27.5′E, 86 m)
  Collector: Yamaguchi et al. (R/V Tansei)
  Composition: $N=47$, $N_R=16$, $N_Q=31$, CV=1 (0R+1Q), RV=21 (10R+11Q), LV=25 (6R+19Q)

Cryptopecten phrygium (Dall)

Confirmed Recent specimens of *C. phrygium**
USNM 62352. Station 45, Mexican Gulf (25°33′N, 84°21′W, 101 fms) $N=1$, CV=1
USNM 694425. Station 1253 of m/v Oregon, Mexican Gulf (29°47′N, 87°17′W, 70 fms) $N=1$, RV=1
USNM 764542. 115 miles off southwest Louisiana (28°02′N, 93°05′W, 47–65 fms) $N=2$, LV=2
USNM 764544. Off Hillsboro Light, Florida (80–100 fms) $N=1$, CV=1

---

* Because *C. phrygium* is not represented in the present collection, the specimens in the National Museum of Natural History, Washington D. C. (USNM) and the American Museum of Natural History, New York (AMNH) were used in this study. Locality and other information have been taken directly from the museum labels.

USNM 764545. 80 miles southwest off Cape San Blas, Florida (Station 50 of Blake Expedition) (26°31′N, 85°53′W, 119 fms) $N=4$, RV=2, LV=2
USNM 765188. Station 10513 of m/v Oregon, ca. 100 miles north of Georgetown, Guiana (8°26′N, 58°11′W, 100 fms) $N=1$, CV=1
AMNH 167207. East edge of Desoto Canyon, west of Cape San Blas, Florida (100–300 fms) (pars) $N=4$, RV=3, LV=1
AMNH 171433. Fowler Light, east coast of Florida $N=1$, CV=1
AMNH 171432. Eolis Station 32, off Sand Key, Dade County, Florida (61 fms) $N=3$, CV=3
AMNH 99641. 110 miles southwest of Egmont Key, Florida (85–100 fms) $N=4$, RV=3, LV=1
AMNH 190947. Dredged, Key West, Florida $N=1$, CV=1
AMNH 99686. 90 miles southwest of Egmont Key, Florida (60 fms) $N=1$, LV=1
AMNH 190960. No locality record $N=1$, RV=1
AMNH 190948. Southwest of Egmont Key, Florida (200 fms) $N=4$, RV=1, LV=3
AMNH 190949. 110 miles southwest of Egmont Key, Florida (90 fms) $N=1$, CV=1

Cryptopecten spinosus sp. nov.
Fossil sample

Sample *Kk* (*S*) [UMUT CM16170]
  Locality: Road-cut, about 500 m north of Kamikatetsu, Kikai Island (Kikai Town), Kagoshima Prefecture [28°17.0′N, 129°56.8′E]
  Horizon: Coral sand of Wan Formation of Ryukyu Group
  Age: Late Pleistocene (0.08 Ma, according to Sakanoue et al., 1967, and Omura, 1983)
  Collector: Yamaguchi and Hayami
  Composition: $N=190$, RV=106, LV=84

Cryptopecten yanagawaensis (Nomura and Zinbo)
Fossil sample

Sample *Mn* (*Y*) [UMUT CM16171]
  Locality: A small valley, southwest of Jyunishin, Natori City, Miyagi Prefecture [38°12.3′N, 140°50.5′E]
  Horizon: Coarse sandstone of lower part of Moniwa Formation of Natori Group
  Age: Early Middle Miocene (N8, 15–16 Ma)
  Collector: Ozawa, Sato and Hayami
  Composition: $N=41$, RV=25, LV=16

Other confirmed fossil specimens of *C. yanagawaensis*

IGPS no. 85962. Moniwa Formation at about 500 m east of Kitaakaishi, Akiu village, Miyagi Prefecture

IGPS no. 90785. Kanagase Member at Yujiri of Kanagase, Ôgawara Town, Miyagi Prefecture

IGPS no. 95010. Sunakozaka Formation at a river cliff of the Asano, Higashiichise, Kanazawa City (described by Ogasawara, 1976)

SM no. 2140, 3545. Moniwa Formation at Moniwa, Sendai City

SM no. 3546. Moniwa Formation about 400 m west of Jyunishin, Kumanodo, Natori City, Miyagi Prefecture

SM no. 8353. Yanagawa Formation at Yanagawa Town, Fukushima Prefecture (type material described by Nomura and Zinbo, 1936)

ACKNOWLEDGMENTS

From its beginnings with the preparation of previous articles, this study has been supported by the generous help of numerous persons and organizations. I would first like to express my sincere thanks to Dr. Thomas R. Waller (Smithsonian Institution) for his kind advice, encouragement and critical reading of the first draft of this manuscript as well as for his hospitality during my stay in Washington, D. C. His unpublished data on the type specimens of early described pectinids in various European institutions, which he was good enough to make available, were particularly instructive for the present study. In the course of this study I received much invaluable information and many suggestions from the following scientists, to whom I am also greatly indebted: Professor Stephen Jay Gould (Harvard University), Professor Richard A. Reyment (Uppsala University), Dr. Motoo Kimura and his colleagues (National Institute of Genetics, Mishima), Professors Tetsuro Hanai and Masuoki Horikoshi (University of Tokyo), Dr. Kiyotaka Chinzei (University of Tokyo), Dr. Tomowo Ozawa (Hyogo Education University), Dr. Toshiyuki Yamaguchi (Chiba University), Dr. Tadashige Habe (Tokai University), Dr. Takashi Okutani (National Science Museum, Tokyo), Dr. Katsura Oyama (Toba Marine Aquarium) and Dr. Henk H. Dijkstra (Sneek, Netherlands). Dr. D. D. Swinbanks (University of Tokyo) kindly read the final draft of this manuscript and gave me many suggestions in English writing. The material studied here was kindly shown, loaned or donated by the following persons, without whose generous cooperation this work could never have been completed: Drs. Norman D. Newell and William K. Emerson (American Museum of Natural History), Dr. Joseph Rosewater (Smithsonian Institution), Dr. Suguru Ohta and Mr. Eiji Tsuchida (University of Tokyo), Dr. Minoru Imajima and Messrs. Akihiko Matsukuma and Tomoki Kase (National Science Museum, Tokyo), Dr. Hiromichi Hirano (Waseda University), the late Professor Tokio Shikama (Yokohama National University), Mr. Yoshiaki Matsushima (Kanagawa Prefectural Museum), Professor Sakae Ohara (Chiba University), Dr. Hiroshi Kitazato (Shizuoka University), Dr. Kenshiro Ogasawara (Tohoku University), Mr. Shoichiro Hayashi (Mikawa-isshiki), Mr. Kazuo Kurohara (Tosa-shimizu), Mr. Atsushi Fujii (Kitakyushu City Museum), Professor Tsugio Shuto and Mr. Yoshio Sato (Kyushu University), Mr. Aizo Yamamoto (Nagasaki), Dr. Hiroyuki Otsuka (Kagoshima University) and Professor Tomohide Nohara (Ryukyu University). I am also much indebted to the staff members of the following institutions for making available various facilities for my study of their collections: Department of Invertebrate Zoology, National Museum of Natural History; Department of Invertebrates, American Museum of Natural History; Institute of Geology and Palaeontology, Tohoku University; Department of Zoology, National Science Museum [Tokyo]; and Geological Institute, Yokohama National University. I thank the captains and crews of R/V Tansei and R/V Rinkai as well as many students of the University of Tokyo for their cooperation in dredging Recent samples, Dr. Michio Shigei and other staff members of the Misaki Marine Biological Station for making available various facilities, and also

Mmes. Reiko Mitsuda and Yoko Suwa and Misses Kasumi Ohma, Masae Toyama and Megumi Ikawa for their patient efforts in the preparation of samples, SEM photographs and this manuscript. Expenses for this study were partly defrayed through Grants-in-aid (Nos. 348023, 56540480) from the Ministry of Education, Science and Culture, Japan.

REFERENCES

Abbott, R. T. (1954)   *American Seashells*. Toronto, Van Nostrand. 541p., 39 pls.

Abbott, R. T. (1974)   *American Seashells*. 2nd ed. New York, Van Nostrand. 663p., 24 pls.

Bavay, A. (1904)   Descriptions de quelques nouvelles espèces du genre *Pecten* et retifications. *Jour. Conch.*, **52**: 197–206, pl. 6.

Bernardi, M. (1858)   Description d'espèces nouvelles. *Jour. Conch.*, **7**: 90–94, pls. 1, 2.

Berry, R. J. and Crothers, J. H. (1974)   Visible variation in the dog-whelk, *Nucella lapillus*. *Jour. Zool. London*, **174**: 123–148.

Beu, A. G. (1966)   Ecological variation of *Chlamys dieffenbachi* (Reeve) (Mollusca, Lamellibranchiata). *Trans. Roy. Soc. New Zealand (Zool.)*, **7**: 93–96.

Biological Laboratory, Imperial Household (1971)   *The Sea Shells of Sagami Bay collected by His Majesty the Emperor of Japan* [described by Kuroda, T., Habe, T. and Oyama, K.]. Tokyo, Maruzen. 1203p., 121 pls. [in Japanese and English]

Brinkmann, R. (1929)   Statistisch-biostratigraphische Untersuchungen an mitteljurassischen Ammoniten über Artbegriff und Stammesentwicklung. *Abh. Gesell. Wiss. Göttingen, Math.-Phys. Kl.*, **13**: 1–249.

Cain, A. J. and Sheppard, P. M. (1950)   Selection in the polymorphic land snail *Cepaea nemoralis*. Heredity, **4**: 275–294.

Chinzei, K. (1978)   Neogene molluscan faunas in the Japanese Islands: an ecologic and zoogeographic synthesis. *Veliger*, **21**: 155–170.

Colton, H. S. (1922)   Variation in the dog whelk *Thais* (*Purpura* auct.) *lapillus*. *Ecology*, **3**: 146–157.

Cox, I. (1957, ed.)   *The Scallop. Studies of a Shell and its Influence on Humankind*. London, Shell Transport and Trading Co., 135p.

Cox, L. R. et al. (1969)   *Treatise on Invertebrate Paleontology*. Part N, Mollusca 6, Bivalvia, vols. 1 and 2, p. N1–N952. Geol. Soc. America and Kansas Univ.

Cox, L. R. (1927)   Report on the palaeontology of the Zanzibar Protectorate. Neogene and Quaternary Mollusca from the Zanzibar Protectorate. pp. 13–102, pls. 3–19.

Cox, L. R. (1930)   Report on geological collections from the coastlands of Kenya Colony made by Miss M. McKinnon Wood. V: Miocene Mollusca, VI: Pliocene Mollusca, VII: Post-Pliocene Mollusca. *Monogr. Geol. Dept. Hunterian Mus., Glasgow Univ.*, **4**: 103–163, pls. 12–15.

Dakin, W. J. (1909)   *Pecten*. Memoirs on typical British marine plants and animals, XVII. Liverpool Marine Biology Committee. 136p., 9 pls.

Dall, W. H. (1886)   Reports on the results of dredging, under the supervision of Alexander Agassiz, in the Gulf of Mexico (1877–78) and in the Caribbean Sea (1879–80), by the U.S. Coast Survey Steamer "Blake", Lieut.-Commander C. D. Sigsbee, U. S. N., and Commander J. R. Bartlett, U. S. N., Commanding. XXIX, Report on the Mollusca,—Part 1. Brachiopoda and Pelecypoda. *Bull. Mus. Comp. Zool., Harvard Coll.*, **12** (6): 171–318, pls. 1–9.

Dall, W. H., Bartsch, P. and Rehder, H. A. (1938)   A manual of the Recent and fossil marine pelecypod molluscs of the Hawaiian Islands. *Bernice P. Bishop Mus., Bull.* **153**: 1–233, pls. 1–58.

Dautschenberg, P. and Bavay, A. (1912)   *Les Lamellibranches de l'Expedition du Siboga*. Partie systematique. 1. Pectinidés. pp. 127–167, pls. 27, 28.

142

Diver, C. (1929) Fossil records of Mendelian mutants. *Nature*, **124**: 183.

Drew, G. A. (1906) The habits, anatomy, and embryology of the giant scallop (*Pecten tenuicostatus*, Mighels). *Maine Univ. Studies*, **6**: 1–71.

Dunker, W. (1877) Mollusca nonnulla nova maris japonici. *Malac. Blätt.*, **24**: 67–75.

Dunker, W. (1882) *Index Molluscorum maris Japonici.* Cassellis Cottorum. Sumptibus Theodori Fischer. 301 p., 16 pls.

Durham, J. W. (1978) Polymorphism in the Pliocene sand dollar *Merriamaster* (Echinoidea). *Jour. Paleontol.*, **52**: 275–286.

Eames, F. E. and Cox, L. R. (1956) Some Tertiary Pectinacea from East Africa, Persia and the Mediterranean rigion. *Proc. Malacol. Soc. London*, **32**: 1–68, pls. 1–20.

Eldredge, N. and Gould, S. J. (1972) Punctuated equilibria: an alternative model to phyletic gradualism. *In* Schopf, T. J. M. (ed.): *Models in Paleobiology*. San Francisco, Freeman. pp. 82–115.

Eldredge, N. and Gould, S. J. (1977) Evolutionary models and biostratigraphic strategies. *In* Kauffman, E. G. and Hazel, J. E. (ed.): *Concepts and Methods of Biostratigraphy*. Stroudsburg, Hutchinson and Ross. pp. 25–40.

Falconer, D. S. (1960) *Introduction to Quantitative Genetics*. Edinburgh, Oliver and Boyd. 365 p.

Fatton, E. (1973) De la province biogeographique à la population d'après les pectinides néogènes et actuels. *Centre d'Etudes et de Recherches de Paléontologie Biostratigraphique, Notes et Contributions*, **8**: 1–97, pls. 1–5.

Ford, E. B. (1940) Polymorphism and taxonomy. *In* Huxley, J. S. (ed.): *The New Systematics*. Oxford, Clarendon Press. pp. 493–513.

Fretter, V. and Graham, A. (1963) The origin of species in littoral prosobranches. *In* Harding, J. P. and Tebble, N. (ed.): *Speciation in the Sea*. System. Assoc. Publ., no. 5, pp. 99–107.

Fujita, T. (1929) Report on the dredged shells of Tateyama Bay (1). *Venus*, **1**: 58–65. [in Japanese]

Goldschmidt, R. (1940) *The Material Basis of Evolution*. Yale Univ. Press. 436 p.

Gould, S. J. (1966) Allometry and size in ontogeny and phylogeny. *Biol. Review*, **41**: 587–640.

Gould, S. J. (1969) An evolutionary microcosm: Pleistocene and Recent history of the land snail *P.* (*Poecilozonites*) in Bermuda. *Bull. Mus. Comp. Zool.*, **138** (7): 407–532.

Gould, S. J. (1971) Muscular mechanics and the ontogeny of swimming in scallops. *Palaeontology*, **14**: 61–94.

Gould, S. J. (1982) Darwinism and the expansion of evolutionary theory. *Science*, **216**: 380–387.

Gould, S. J. and Eldredge, N. (1977) Punctuated equilibria: the tempo and mode of evolution reconsidered. *Paleobiology*, **3**: 115–151.

Gould, S. J. and Johnson, R. F. (1972) Geographic variation. *Ann. Rev. Ecol. System.*, **3**: 457–498.

Grau, G. (1959) Pectinidae of the eastern Pacific. *Allan Hancock Pacific Expeditions*, **23**: 1–308, pls. 1–57.

Habe, T. (1951) *Genera of Japanese Shells*. Pelecypoda, 1. Tokyo, Kairui-Bunken-Kankokai. 96 p. [in Japanese]

Habe, T. (1958) Report on the Mollusca chiefly collected by the S. S. Soyo-maru of the Imperial Fisheries Experimental Station on the continental shelf bordering Japan during the years 1922–1930. Part 3 and 4. *Publ. Seto Mar. Biol. Lab.*, **6**: 241–280, pls. 11–13; **7**: 19–52, pls. 1, 2.

Habe, T. (1961) *Colored Illustrations of the Shells of Japan (II)*. Osaka, Hoikusha. 223 p., 66 pls. [in Japanese]

Habe, T. (1964)   *Shells of the Western Pacific in Color.* vol. 2. Osaka, Hoikusha. 223 p., 66 pls. [English edition of Habe (1961)]

Habe, T. (1977)   *Systematics of Mollusca in Japan.* Bivalvia and Scaphopoda. Tokyo, Hokuryukan. 372 p. [in Japanese]

Habe, T. and Kosuge, S. (1967) Shells. Osaka, Hoikusha. 223 p., 64 pls. [in Japanese]

Hallam, A. (1975)   Evolutionary size increase and longevity in Jurassic bivalves and ammonites. *Nature,* **258**: 493–496.

Hayami, I. (1972)   Possibility of "paleogenetics". *Jour. Geol. Soc. Japan,* **78**: 495–506. [in Japanese with English abstract]

Hayami, I. (1973)   Discontinuous variation in an evolutionary species, *Cryptopecten vesiculosus,* from Japan. *Jour. Paleontol.,* **47**: 401–420, pls. 1, 2.

Hayami, I. (1982)   Taxonomic names of *Cryptopecten* species. *Venus,* **41**: 233–236. [in Japanese with English abstract]

Hayami, I. and Matsukuma, A. (1970)   Variation of bivariate characters from the standpoint of allometry. *Palaeontology,* **13**: 588–605.

Hayami, I. and Matsukuma, A. (1971)   Mensuration of fossils and statistics—Analysis of allometry and variation. *Sci. Rep. Dept. Geol. Kyushu Univ.,* **10**: 135–160. [in Japanese with English abstract]

Hayami, I. and Ozawa, T. (1975)   Evolutionary models of lineage-zones. *Lethaia,* **8**: 1–14.

Hayasaka, S. (1973)   Pliocene marine fauna from Tane-ga-shima, south Kyushu, Japan. *Sci. Rept. Tohoku Univ. Sendai,* [2], Spec. Vol. 6: 97–108, pls. 6, 7.

Hedley, F. L. S. (1909)   Mollusca from the Hope Islands, north Queensland. *Proc. Linn. Soc. New South Wales,* **34**: 420–466, pls. 36–44.

Hirano, H. (1978)   Phenotypic substitution of *Gaudryceras* (a Cretaceous ammonite). *Trans. Proc. Palaeont. Soc. Japan,* [N. S.], **109**: 235–258, pls. 33–35.

Hirase, S. (1934)   *A Collection of Japanese Shells with Illustrations in Natural Colors.* Tokyo, Matsumura-Sanshodo. 217 p., 129 pls. [in Japanese and English]

Horikoshi, M. (1957)   Note on the molluscan fauna of Sagami Bay and its adjacent waters. *Sci. Rept. Yokohama Nat. Univ.,* ser. 2, **6**: 37–64.

Horikoshi, M. (1960)   A topographical approach to a study of the benthic communities on the submarine ridge, Sé-no-umi, in Sagami Bay. *Sci. Rep. Yokosuka City Mus.,* **5**: 6–8.

Horikoshi, M. et al. (1982)   *Priliminary Compilation of the Results obtained by the "Survey of Continental Shelf bordering Japan" carried out on board the S/S Soyo-maru during 1923–1930.* Report for a part of the result of Grant-in-aid for Co-operative Research (A), Ministry of Education, Science and Culture: "Studies on off-shore and deep-sea faunas in the West Pacific and Indian Oceans." Tokyo. 252 p.

Huxley, J. S. (1924)   Constant differential growth-ratios and their significance. *Nature,* **114**: 895–896.

Huxley, J. S. (1932)   *Problem of Relative Growth.* London, Methuen. 312 p.

Imajima, M. (1970)   Errant polychaetous annelids collected from the areas around the Tsushima Islands. *Mem. Nat. Sci. Mus.* [*Tokyo*], **3**: 113–122.

Imbrie, J. (1956)   Biometrical methods in the study of invertebrate fossils. *Bull. Amer. Mus. Nat. Hist.,* **108**: 217–252.

Iredale, T. (1939)   Great Barrier Reef expedition 1928–1929. Mollusca, pt. 1. *Brit. Mus.* (*Nat. Hist.*), *Sci. Rept.* **5** (6): 209–425, pls. 1–7.

Itoigawa, J., Shibata, H. and Nishimoto, H. (1974)   Molluscan fossils of the Mizunami Group. *Bull. Mizunami Fossil Mus.,* **1**: 43–203, pls. 1–63.

Itoigawa, J., Shibata, H., Nishimoto, H. and Okumura, Y. (1981)   Miocene fossils of the Mizunami Group, central Japan. 2. Molluscs. *Monogr. Mizunami Fossil Mus.,* **3-A**: 1–53, pls.

144

1–52.

Itoigawa, J., Shibata, H., Nishimoto, H. and Okumura, Y. (1982) Miocene fossils of the Mizunami Group, central Japan. 2. Molluscs. *Monogr. Mizunami Fossil Mus.*, **3-B**: 1–330.

Johnson, A. L. A. (1981) Detection of ecophenotypic variation in fossils and its application to a Jurassic scallop. *Lethaia*, **14**: 277–285.

Kanno, S. (1958) New Tertiary Mollusca from the Chichibu Basin, Saitama Prefecture, central Japan. *Sci. Rept. Tokyo Kyoiku Daigaku*, [C] **6**: 157–229, pls. 1–7.

Kanno, S. (1960) *The Tertiary System of the Chichibu Basin, Saitama Prefecture, Central Japan.* Part 2. Palaeontology. Tokyo, Japan Soc. Prom. Sci., pp. 123–396, pls. 31–51.

Kay, E. A. (1979) Hawaiian marine shells. Reef and shore fauna of Hawaii, Section 4: Mollusca. *Bernice P. Bishop Mus., Spec. Publ.* **64** (4): 1–653.

Keigwin, L. D. J. (1978) Pliocene closing of the Isthmus of Panama, based on biostratigraphic evidence from nearby Pacific Ocean and Caribbean Sea cores. *Geology*, **6**: 630–634.

Kellogg, D. E. (1975) The role of phyletic change in the evolution of *Pseudocubus vema* (Radiolaria). *Paleobiology*, **1**: 359–370.

Kermack, K. A. and Haldane, J. B. S. (1950) Organic correlation and allometry. *Biometrika*, **37**: 30–41.

Kettlewell, H. B. D. (1961) The phenomenon of industrial melanism in Lepidoptera. *Ann. Rev. Entomol.*, **6**: 245–262.

Kimura, M. (1960) *Outline of Population Genetics.* Tokyo, Baifukan. 312 p. [in Japanese]

Kira, T. (1950) "Different morphs in shells". *Yumehamaguri*, **6**: 49. [in Japanese]

Kira, T. (1954) *Colored Illustrations of the Shells of Japan.* Osaka, Hoikusha. 209 p., 68 pls. [in Japanese]

Kira, T. (1962) *Shells of the Western Pacific in Color.* vol. 1. Osaka Hoikusha. 224 p., 68 pls. [English edition of Kira (1954)]

Knox, G. A. (1963) Problems of speciation in intertidal animals with special reference to New Zealand shores. *System. Assoc. Publ.*, **5**: 7–29.

Komai, T. (1956) Genetics of ladybeetles. *Advance. in Genetics*, **8**: 155–188.

Komai, T. and Emura, S. (1955) A study of population genetics of the polymorphic land snail *Bradybaena similaris. Evolution*, **9**: 400–418.

Konishi, K., Schlanger, S. O. and Omura, A. (1970) Neotectonic rates in the central Ryukyu Islands derived from ^{230}Th coral ages. *Marine Geology*, **9**: 225–240.

Kuroda, T. (1932) An illustrated catalogue of the Japanese shells (10). *Venus*, **3**: appendix 87–102. [in Japanese]

Kuroda, T. and Habe, T. (1952) *Check List and Bibliography of the Recent Marine Mollusca of Japan.* Tokyo, Hosokawa Print. Co., 210 p.

Kurtén, B. (1955) Contribution to the history of a mutation during 1,000,000 years. *Evolution*, **9**: 107–118.

Küster, H. C. and Kobelt, W. (1888) Die Gattungen *Spondylus* und *Pecten. Syst. Conch. Cab.*, **7**: 1–296, pls. 1–72.

Ladd, H. S. (1945) Mollusca. *In* Ladd, H. S. and Hoffmeister, J. E.: Geology of Lau, Fiji. *Bernice P. Bishop Mus., Bull.* **181**: 331–370, pls. 44–53.

Lamotte, M. (1959) Polymorphism of natural populations of *Cepaea nemoralis. Cold Spring Harbor Symp. Quant. Biol.*, **24**: 65–86.

MacArthur, R. H. (1972) *Geographical Ecology.* Patterns in the Distribution of Species. New York, Harper & Row, 269 p.

Machida, H., Arai, F., Murata, A. and Hakamata, K. (1974) Correlation and chronology of the Middle Pleistocene tephra layers in south Kanto. *Jour. Geogr.* [Tokyo], **83**: 302–338. [in Japanese with English abstract]

Machida, H. and Suzuki, M. (1971) Absolute age of volcanic ashes and the chronology of Late Quaternary. *Kagaku*, **41**: 263–270. [in Japanese]

Makiyama, J. (1931) Stratigraphy of the Kakegawa Pliocene in Totomi. *Mem. Coll. Sci. Kyoto Imp. Univ.*, [B] **7**: 1–53, pl. 1.

Malmgren, B. A. and Kennett, J. P. (1980) Phyletic gradualism in a Late Cenozoic planktonic foraminiferal lineage; DSDP Site 284, southwest Pacific. *Paleobiology*, **7**: 230–240.

Mason, J. (1957) The age and growth of the scallop, *Pecten maximum* (L.), in Manx waters. *Jour. Mar. Biol. Ass. U.K.*, **36**: 473–492.

Masuda, K. (1958) On the Miocene Pectinidae from the environs of Sendai; Part 10, On *Pecten (Aequipecten) yanagawaensis* Nomura and Zinbo. *Trans. Proc. Palaeont. Soc. Japan*, [N. S.] **30**: 189–192, pl. 27.

Masuda, K. (1959) On the Miocene Pectinidae from the environs of Sendai; Part 15, *Pecten cosibensis* Yokoyama and its related species. *Trans. Proc. Palaeont. Soc. Japan*, [N. S.] **35**: 121–132, pl. 13.

Masuda, K. (1962) Tertiary Pectinidae of Japan. *Sci. Rept. Tohoku Univ. Sendai*, [2] **33**: 117–238, pls. 18–27.

Masuda, K. (1966) Molluscan fauna of the Higashi-innai Formation of Noto Peninsula, Japan. II. Remarks on molluscan assemblage and description of species. *Trans. Proc. Palaeont. Soc. Japan*, [N. S.] **64**: 317–337, pls. 34, 36.

Masuda, K. (1973) Cenozoic pectinid fossils from Japan. *Atlas of Japanese Fossils*, no. 33, 24 pp. (6 pls incl.), Tokyo, Tsukiji Shokan. [in Japanese]

Masuda, K. and Noda, H. (1977) *Check List and Bibliography of the Tertiary and Quaternary Mollusca of Japan, 1950–1974*. Sendai, Saito Ho-on Kai. 494 p.

Masuda, K. and Takegawa, H. (1965) Remarks on the Miocene Mollusca from the Sennan district, Miyagi Prefecture, northeast Honshu, Japan. *Saito Ho-on Kai Mus., Res. Bull.*, **34**: 1–14, pls. 1, 2.

Mayr, E. (1942) *Systematics and the Origin of Species*. Columbia Univ. Press. 334 p.

Mayr, E. (1954) Geographic speciation in tropical echinoids. *Evolution*, **8**: 1–18.

Mayr, E. (1963) *Animal Species and Evolution*. Harvard Univ. Press. 797 p.

Mayr, E. (1969) *Principles of Systematic Zoology*. New York, McGraw-Hill. 428 p.

Melvill, J. C. (1888) Description of six new species of *Pecten*. *Jour. Conch.*, **5**: 279–281, pl. 2.

Menzel, R. W. (1968) Chromosome number in nine families of marine pelecypod molluscs. *Nautilus*, **82**: 45–58.

Mitsunashi, T. et al. (1979) Explanatory text of the geological map of Tokyo Bay and adjacent areas. *Geol. Surv. Japan, Misc. Map Series*, **20**: 1–91. [in Japanese with English abstract]

Moore, H. B. (1936) The biology of *Purpura lapillus*. Pt. 1. Shell variation in relation to environment. *Jour. Mar. Biol. Ass. U. K.*, **21**: 61–89.

Mori, S. and Osada, T. (1979) "Catalogue of molluscan fossils from the Ninimiya Formation." *Mater. Rept. Hiratsuka City Mus.*, **19**: 1–42, pls. 1–15. [in Japanese]

Nagao, T. (1928) Paleogene fossils of the islands of Kyushu, Japan, Part 2. *Sci. Rept. Tohoku Imp. Univ. Sendai*, [2] **12**: 11–140, pls. 1–17.

Nemoto, O. and Ohara, S. (1979) "Fossil shells from the Taga Group in the environs of Futatsunuma, Hirono Town." Memorial Volume for the retirement of Prof. I. Yanagisawa. Earth Science Club of Taira. pp. 52–66 (pls. 1–4 incl.). [in Japanese]

Newell, N. D. (1937–1938) Late Paleozoic pelecypods: Pectinacea. *Kansas State Geol. Surv., Bull.* **10** (1): 1–123 [1937], pls. 1–20 [1938].

Newell, N. D. (1956) Fossil Populations. *System. Assoc. Publ.*, **2**: 63–82.

Niino, H. (1935) On the bottom nature of the banks Zenisu, Izu Island. *Jour. Geogr. (Tokyo)*, **47**: 32–37.

146

Niino, H. (1952) On the bottom characters of the insular shelf around Hachijo Island and the neighbouring banks. *Jour. Tokyo Univ. Fish.*, **39**: 101–110, pls. 1, 2.

Noda, H. (1980) Molluscan fossils from the Ryukyu Islands, southwestern Japan. Part 1. Gastropoda and Pelecypoda from the Shinzato Formation in southeastern part of Okinawa-jima. *Sci. Rept. Inst. Geosci. Univ. Tsukuba*, [B] **1**: 1–95, pls. 1–12.

Nomura, E. (1926) An application of $a = kb^x$ in expressing the growth relation in the freshwater bivalve *Sphaerium heterodon* Pils. *Sci. Rept. Tohoku Imp. Univ. Sendai*, [4] **2**: 57–62.

Nomura, S. (1933) Catalogue of the Tertiary and Quartery Mollusca from the Island of Taiwan (Formosa) in the Institute of Geology and Palaeontology, Tohoku Imperial University. *Sci. Rept. Tohoku Imp. Univ. Sendai*, [2] **16**: 1–108, pls. 1–4.

Nomura, S. (1940) Molluscan fauna of the Moniwa shell beds exposed along the Natori-gawa in the vicinity of Sendai, Miyagi Prefecture. *Sci. Rept. Tohoku Imp. Univ. Sendai*, [2] **21**: 1–46, pls. 1–3.

Nomura, S. and Zinbo, N. (1934) Marine Mollusca from the "Ryukyu Limestone" of Kikai-zima, Ryukyu Group. *Sci. Rept. Tohoku Imp. Univ. Sendai*, [2] **16**: 109–164, pl. 5.

Nomura, S. and Zinbo, N. (1936) Additional fossil Mollusca from the Yanagawa shell-beds in the Hukusima basin, northeast Honsyu, Japan. *Saito Ho-on Kai Mus., Res. Bull.*, **10**: 335–345, pl. 20.

Ockelmann, K. W. (1965) Developmental types in marine bivalves and their distribution along the Atlantic coast of Europe. *In* Cox, L. R. and Peake, J. F. (ed.): *Proceedings of the First European Malacological Congress.* pp. 25–35.

Ogasawara, K. (1976) Miocene Mollusca from Ishikawa-Toyama area, Japan. *Sci. Rept. Tohoku Univ.*, [2] **46**: 33–78, pls. 11–15.

Ohara, S. (1968a) The type Semata Formation. *Jour. Coll. Arts and Sci., Chiba Univ., Nat. Sci.*, **5**: 303–318. [in Japanese with English abstract]

Ohara, S. (1968b) *Geological Atlas of the Chiba Prefecture.* No. 5 Bivalvia. Educ. Comm. Chiba Pref. 52 p., 17 pls. [in Japanese]

Ohara, S. (1973) Molluscan fossils from the Higashiyatsu Formation. *Jour. General Educ. Chiba Univ.*, **B-6**: 67–83, pls. 1, 2. [in Japanese with English abstract]

Ohara, S. and Takahashi, Y. (1975) Molluscan fossils of the Kurotaki Formation and pyroclastic rocks of the Anno Formation in the Boso Peninsula (preliminary report). *Jour. General Educ. Chiba Univ.*, **B-8**: 115–129. [in Japanese with English abstract]

Okutani, T. (1972) Molluscan fauna on the submarine banks Zenisu, Hyotanse, and Takase, near the Izu-shichito Islands. *Bull. Tokai Region. Fish. Res. Lab.*, **72**: 63–142, pls. 1, 2.

Okutani, T. and Habe, T. (1975) *Mollusca II. chiefly from Japan.* Tokyo, Gakken. 294 p. (160 pls. incl.). [in Japanese]

Omura, A. (1983) Uranium-series ages of some solitary corals from the Riukiu Limestone on the Kikai-jima, Ryukyu Islands. *Trans. Proc. Palaeont. Soc. Japan*, [N. S.]

Oyama, K. (1973) Revision of Matajiro Yokoyama's type Mollusca from the Tertiary and Quaternary of the Kanto area. *Palaeont. Soc. Japan, Spec. Papers*, **17**: 1–148, pls. 1–57.

Ozawa, T. (1975) Evolution of *Lepidolina multiseptata* (Permian foraminifer) in east Asia. *Mem. Fac. Sci. Kyushu Univ.*, [D] **23**: 117–164, pls. 22–26.

Patterson, C. M. (1969) Chromosomes of molluscs. Proceedings of the symposium on Mollusca held at Cochin from January 12 to 16, 1968. Part 2 (Symposium series 3). Mar. Biol. Assoc. India. pp. 635–686.

Pilsbry, H. A. (1905) New Japanese Mollusca. *Proc. Acad. Nat. Sci. Phila.*, **56**: 550–561, pls. 39–41.

Poutiers, J. M. (1981) Résultats des campagnes MUSORSTOM I—Philippines (18–28 Mars 1976). 16. Mollusques: Bivalves. *Mém. O.R.S.T.O.M.*, **91**: 325–356.

Raup, D. M. (1977)   Stochastic models in evolutionary palaeontology. *In* Hallam, A. (ed.): *Patterns of Evolution, as Illustrated by the Fossil Record.* Amsterdam, Elsevier. pp. 59–78.

Raup, D. M. and Crick, R. E. (1981)   Evolution of single characters in the Jurassic ammonite *Kosmoceras. Paleobiology,* **7**: 200–215.

Raup, D. M. and Crick, R. E. (1982)   *Kosmoceras*: evolutionary jumps and sedimentary breaks. *Paleobiology,* **8**: 90–100.

Raup, D. M., Gould, S. J., Schopf, T. J. M. and Simberloff, D. S. (1973)   Stochastic models of phylogeny and the evolution of diversity. *Jour. Geol.,* **81**: 525–542.

Reeve, L. A. (1853a)   Monograph of the genus *Pecten. Conch. Icon.,* **8**: 1–176, pls. 1–35.

Reeve, L. A. (1853b)   Errata for monograph of the genus *Pecten. Conch. Icon.,* **8**: errata.

Reyment, R. A. (1975)   Analysis of a generic level transition in Cretaceous ammonites. *Evolution,* **28**: 665–676.

Reyment, R. A. (1980)   *Morphometric Methods in Biostratigraphy.* London, Academic Press. 175 p.

Sakanoue, M., Konishi, K. and Komura, K. (1967)   Stepwise determinations of thorium, protactinium and uranium isotopes and their applications in geochronological studies. *In Radioactive Dating and Methods of Low-level Counting.* Intern. Atomic Energy Agency, Vienna. pp. 313–329.

Sheppard, P. M. (1951)   Fluctuations in the selective value of certain phenotypes in the polymorphic land snail *Cepaea nemoralis* (L.). *Heredity,* **5**: 125–134.

Sheppard, P. M. (1952)   Natural selection in two colonies of the polymorphic land snail *Cepaea nemoralis. Heredity,* **6**: 233–238.

Shikama, T. (1964)   *Selected Shells of the World Illustrated in Colours [II].* Tokyo, Hokuryukan. 212 p., 70 pls. [in Japanese]

Shikama, T. (1973)   Molluscan assemblages of the basal part of the Zushi Formation in the Miura Peninsula. *Sci. Rept. Tohoku Univ. Sendai,* [2] Spec. Vol. 6: 179–204, pls. 16, 17.

Shikama, T. and Masujima, A. (1969)   Quantitative studies of the molluscan assemblages in the Ikego-Nojima Formations. *Sci. Rept. Yokohama Nat. Univ.,* [2] **15**: 61–94, pls. 5–7.

Shirai, S. (1958)   On some species of Mollusca collected on the banks of Izu-shichito. *Venus,* **20**: 87–96. [in Japanese]

Shuto, T. (1960)   On some pectinids and venerids from the Miyazaki Group (Palaeontological study of the Miyazaki Group—VII). *Mem. Fac. Sci. Kyushu Univ.,* [D] **9**: 119–149, pls. 12–14.

Simpson, G. G. (1953)   *The Major Features of Evolution.* Columbia Univ. Press. 434 p.

Simpson, G. G., Roe, A. and Lewontin, R. C. (1960)   *Quantitative Zoology* (revised edition). Harcourt, Brace. 440 p.

Sowerby, G. B. (1887)   Monograph of the genus *Pecten. Thes. Conch.,* **1**: 45–82, pls. 12–20.

Sowerby, G. B. (1908)   Description of eight new species of marine Mollusca. *Proc. Malacol. Soc. London,* **8**: 16–19, pl. 1.

Stanley, S. M. (1970)   Relation of shell form to life habits of the Bivalvia (Mollusca). *Geol. Soc. America, Mem.,* **125**: 1–496.

Stanley, S. M. (1979a)   *Macroevolution: Pattern and Process.* San Francisco, Freeman. 332 p.

Stanley, S. M. (1979b)   Evolution. *In* Fairbridge, R. W. and Jablonski, D. (ed.): *The Encyclopedia of Paleontology.* Stroudsburg, Dowden, Hutchinson and Ross. pp. 296–300.

Stanley, S. M., Addicott, W. O. and Chinzei, K. (1980)   Lyellian curves in paleontology: possibilities and limitations. *Geology,* **8**: 422–426.

Stenzel, H. B. (1971)   *Treatise on Invertebrate Paleontology.* Part N, Mollusca 6, Bivalvia, vol. 3. Oysters. pp. N953–N1224. Geol. Soc. America and Kansas Univ.

Sugihara, S., Arai, F. and Machida, H. (1978)   Tephrochronology of the Middle to Late Pleistocene sediments in the northern part of the Boso Peninsula, central Japan. *Jour. Geol.*

Soc. Japan, **84**: 583–600. [in Japanese with English abstract]

Taki, I. and Oyama, K. (1954)   Matajiro Yokoyama's the Pliocene and later faunas from the Kwanto region in Japan. *Palaeont. Soc. Japan, Spec. Papers,* **2**: 1–68, pls. 1–49.

Thayer, C. W. (1972)   Adaptive features of swimming monomyarian bivalves (Mollusca). *Forma Functio,* **5**: 1–31.

Tsuchi, R. (1981, ed.)   *Neogene of Japan—Its Biostratigraphy and Chronology.* IGCP-114 National Working Group of Japan, Shizuoka. 140 p.

Vokes, H. E. (1967)   Genera of the Bivalvia: a systematic and bibliographic catalogue. *Bull. Amer. Paleont.,* **51**: 105–394.

Vokes, H. E. (1980)   *Genera of the Bivalvia: a systematic and bibliographic catalogue (revised and updated).* Paleont. Res. Inst. Ithaca, N. Y., 307 p.

Waller, T. R. (1969)   The evolution of the *Argopecten gibbus* stock (Mollusca: Bivalvia), with emphasis on the Tertiary and Quaternary species of eastern North America. *Paleont Soc., Mem.* **3**: 1–125 (pls. 1–8 incl.).

Waller, T. R. (1972a)   The Pectinidae (Mollusca Bivalvia) of Eniwetok Atoll, Marshall Islands. *Veliger,* **14**: 221–264, pls. 1–8.

Waller, T. R. (1972b)   The functional significance of some shell microstructure in the Pectinacea (Mollusca: Bivalvia). *24th Intern. Geol. Congress,* section **7**: 48–56.

Waller, T. R. (1973)   The habits and habitats of some Bermudian marine mollusks. *Nautilus,* **87**: 31–52.

Waller, T. R. (1978)   Morphology, morphoclines and a new classification of the Pteriomorphia (Mollusca: Bivalvia). *Phil. Trans. Roy. Soc. London,* [B] **284**: 345–365.

Waller, T. R. (1981)   Functional morphology and development of veliger larvae of the European oyster, *Ostrea edulis* Linné. *Smiths. Contr. to Zool.,* **328**: 1–70.

Williamson, P. G. (1981)   Paleontological documentation of speciation in Cenozoic molluscs from Turkana Basin. *Nature,* **293**: 437–443.

Wilson, E. C. (1975)   Light show from beyond the grave. *Terra* [Members Magazine, Natural History Museum of Los Angeles County], **13**: 10–13.

Woodring, W. P. (1982)   Geology and paleontology of Canal Zone and adjoining parts of Panama. Description of Tertiary molluscs. *U. S. Geol. Surv. Prof. Paper,* **306** (F): 541–759, pls. 83–124.

Yokoyama, M. (1911)   Pectens from the Koshiba Neogene. *Jour. Geol. Soc. Tokyo,* **18**: 1–5, pl. 1.

Yokoyama, M. (1920)   Fossils from the Miura Peninsula and its immediate north. *Jour. Coll. Sci. Imp. Univ. Tokyo,* **36**: 1–186, pls. 1–20.

Yokoyama, M. (1922)   Fossils from the Upper Musashino of Kazusa and Shimosa. *Jour. Coll. Sci. Imp. Univ. Tokyo,* **44**: 1–200, pls. 1–17.

Yokoyama, M. (1925)   Molluscan remains from the uppermost part of the Jō-ban coal-field. *Jour. Coll. Sci. Imp. Univ. Tokyo,* **45**: 1–34, pls. 1–6.

Yonge, C. M. (1936)   The evolution of the swimming habit in the Lamellibranchia. *Mem. Mus. Roy. d'Hist. Nat. Belge,* **3**: 77–100.

Yoshida, H. (1964)   *Seeding of shell-fish.* Tokyo, Hokuryukan. 221 p. [in Japanese]

Yoshihara, S. (1902)   Japanese shells (lamellibranchs). *Zool. Mag. Tokyo,* **14**: 141–145, 208–212, 356–358, pls. 1–5. [in Japanese]

POSTSCRIPT

After the manuscript of this paper was submitted, I read W. Zhenrui's (1983) article on the systematics of *Cryptopecten* from Chinese waters. The following four species were described with indication of some sampling localities: *Cryptopecten complanus* sp. nov., *C. tissotii* [=*C. bullatus* in the present paper], *C. nux* and *C. vesiculosus*.

C. complanus was based on two specimens from East China Sea; the holotype (Reg. no. M11072, Institute of Oceanology, Academia Sinica, Qingdao) at 128°E, 31°5′N (147 m) and the paratype at 126°E, 27°5′N (137 m). As was discussed by Zhenrui, this species seems to be closely related to *C. bullatus* in view of the similar number of radial ribs and similarly alternate imbricated scales. It appears to have been considered by him that *C. bullatus* may be distinguished from the new species by the stronger convexity of two valves, the presence of a serrated riblet on either side of the rib, and somewhat deeper byssal notch. Because these diagnostic characters do not seem to be very conclusive, I wish to give my opinion about the taxonomic validity of *C. complanus* after observing its type specimens or their photographs.

The occurrence of *C. bullatus* and *C. nux*, the specific assignment of which is confirmed through Zhenrui's description and illustration, were recorded from South China Sea between the mouth of the River Si and the Paracel Islands at the depth of 104 m and 39–41 m, respectively. *C. vesiculosus* was said to be distributed in East China Sea from Cheju (Quelpart) Island on the north to Taiwan Strait on the south (82–150 m). These new locality records seem to be consistent with the ranges of geographic and bathymetric distribution described in the present paper.

Finally, I thank Dr. Takashi Okutani for his prompt information on this new article.

Reference

Zhenrui, W. (1983) Studies on Chinese species of the family Pectinidae. III. Chlamydinae (a new species and three new records of the genus *Cryptopecten*). *Oceanologia et Limnologia Sinica*, **14**: 402–406.

PLATES

Explanation of Plate 1

(All figures ×2)

Figs. 1–6.　*Cryptopecten bullatus* (Dautzenberg and Bavay)

Fig. 1.　Right valve (USNM 764151), off Waikiki, Hawaii.

Fig. 2.　Right valve (USNM 764151), the same locality; 2a: external view, 2b: internal view, 2c: anterior view, 2d: dorsal view.

Fig. 3.　Left valve (USNM 764151), the same locality; 3a: external view, 3b: internal view.

Fig. 4.　Left valve (USNM 294674), off N. Burias, Philippines; 4a: external view, 4b: internal view.

Fig. 5.　Living conjoined valves (UMUT RM16002b), sample *Hs* (*3*) (*B*), Hyôtanse near Niijima Island; 5a: external view of right valve, 5b: internal view of right valve, 5c: external view of left valve, 5d: internal view of left valve, 5e: anterior view, 5f: dorsal view.

Fig. 6.　Living conjoined valves (USNM 764152), Tosa Bay, Shikoku; 6a: external view of right valve, 6b: internal view of right valve, 6c: external view of left valve, 6d: internal view of left valve, 6e: anterior view, 6f: dorsal view.

1
2a
2b
2c
2d
3a
3b
4a
4b
5a
5b
5e
6a
5c
5d
5f
6b
6c
6d
6e
6f

Explanation of Plate 2

(All figures ×2, unless otherwise stated)

Figs. 1–3. *Cryptopecten bullatus* (Dautzenberg and Bavay)

Fig. 1. Right valve (UMUT CM16001a), sample *Kk* (*B*), Late Pleistocene Wan Formation at Kikai Island; la: external view, 1b: internal view.

Fig. 2. Left valve (UMUT CM16001b), the same sample; 2a: external view, 2b: internal view.

Fig. 3. Living conjoined valves (UMUT RM16007a), sample *Ks* (*B*), mouth of Kii Strait; 3a: external view of right valve, 3b: external view of left valve, 3c: anterior view, 3d: dorsal view, 3e: ornament on ventral area of right valve showing alternate disposition of imbricated scales (×5, with whitening), 3f: byssal wing (×5, with whitening).

Fig. 4. *Cryptopecten nux nux* (Reeve)

Fig. 4. Living conjoined valves (UMUT RM16020), sample *An* (*N*), mouth of Ise Bay; 4a: external view of right valve, 4b: internal view of right valve, 4c: external view of left valve, 4d: internal view of left valve, 4e: anterior view, 4f: dorsal view, 4g: ornament on ventral area of right valve showing alternate disposition of almost exfoliated imbricated scales (×5, with whitening), 4h: byssal wing (×5, with whitening)

1a
1b
2a
2b
3a
3b
3c
3d
4g
3e
4a
4b
4e
3f
4c
4d
4f
4h

Explanation of Plate 3

(All figures × 2, unless otherwise stated)

Figs. 1–2. *Cryptopecten nux nux* (Reeve)

    Fig. 1.   Right valve (UMUT CM16015c), sample *Kk* (*N*), Late Pleistocene Wan Formation at Kikai Island; 1a: external view, 1b: internal view, 1c: anterior view, 1d: dorsal view.

    Fig. 2.   Left valve (UMUT CM16015d), the same sample; 2a: external view, 2b: internal view, 2c: anterior view, 2d: dorsal view.

Fig. 3–4. *Cryptopecten nux sematensis* (Oyama)

    Fig. 3.   Right valve (UMUT CM21561), Holotype, Yokoyama's specimen (1922, pl. 15, fig. 1), "Shito" (probably identical with the locality of sample *Sm* (*N*)); 3a: external view, 3b: internal view, 3c: anterior view, 3d: dorsal view.

    Fig. 4.   Left valve (UMUT CM16180a), sample *Sm* (*N*), Late Pleistocene Yabu Formation at Ochishimoshinden; 4a: external view, 4b: internal view, 4c: anterior view, 4d: dorsal view.

Fig. 5. *Cryptopecten vesiculosus vesiculosus* (Dunker)

    Fig. 5.   Population sample (UMUT RM16138), sample *Is*, mouth of Ise Bay; Left: individuals belonging to Phenotype *R*, Right: individuals belonging to Phenotype *Q*. (× 0.8)

1a
1b
1c
1d
2a
2b
2c
2d
3a
3b
3c
3d
4a
4b
4c
4d
5

Explanation of Plate 4

(All figures ×1.5)

Figs. 1–3. *Cryptopecten vesiculosus vesiculosus* (Dunker)

Fig. 1. Living conjoined valves (UMUT RM16138b), Phenotype *Q* (*rough* subphenotype), sample *Is*, mouth of Ise Bay; 1a: external view of right valve, 1b: internal view of right valve, 1c: external view of left valve, 1d: internal view of left valve.

Fig. 2. Living conjoined valves (UMUT RM16138c), Phenotype *Q* (*smooth* subphenotype), the same sample; 2a: external view of right valve, 2b: internal view of right valve, 2c: external view of left valve, 2d: internal view of left valve.

Fig. 3. Living conjoined valves (UMUT RM16138d), Phenotype *R*, the same sample; 3a: external view of right valve, 3b: internal view of right valve, 3c: external view of left valve, 3d: internal view of left valve.

1a
2a
3a
1b
2b
3b
1c
2c
3c
1d
2d
3d

Explanation of Plate 5

(×1.5 for Figs. 1a–c, 2a, b, 3a–c, ×4.5 for Figs. 1d–f, 2c, d, 3d–f)

Figs. 1–3. *Cryptopecten vesiculosus vesiculosus* (Dunker)

Fig. 1. Living conjoined valves (UMUT RM16138b), the same specimen as Pl. 4, Fig. 1; 1a: anterior view, 1b: dorsal view, 1c: ventral view, 1d: byssal wing (with whitening), 1e: ventral area of right valve (with whitening), 1f: ventral area of left valve (with whitening).

Fig. 2. Living conjoined valves (UMUT RM16138c), the same specimen as Pl. 4, Fig. 2; 2a: anterior view, 2b: dorsal view, 2c: ventral area of right valve (with whitening), 2d: ventral area of left valve (with whitening).

Fig. 3. Living conjoined valves (UMUT RM16138d), the same specimen as Pl. 4, Fig. 3; 3a: anterior view, 3b: dorsal view, 3c: ventral view, 3d: byssal wing (with whitening), 3e: ventral area of right valve (with whitening), 3f: ventral area of left valve (with whitening).

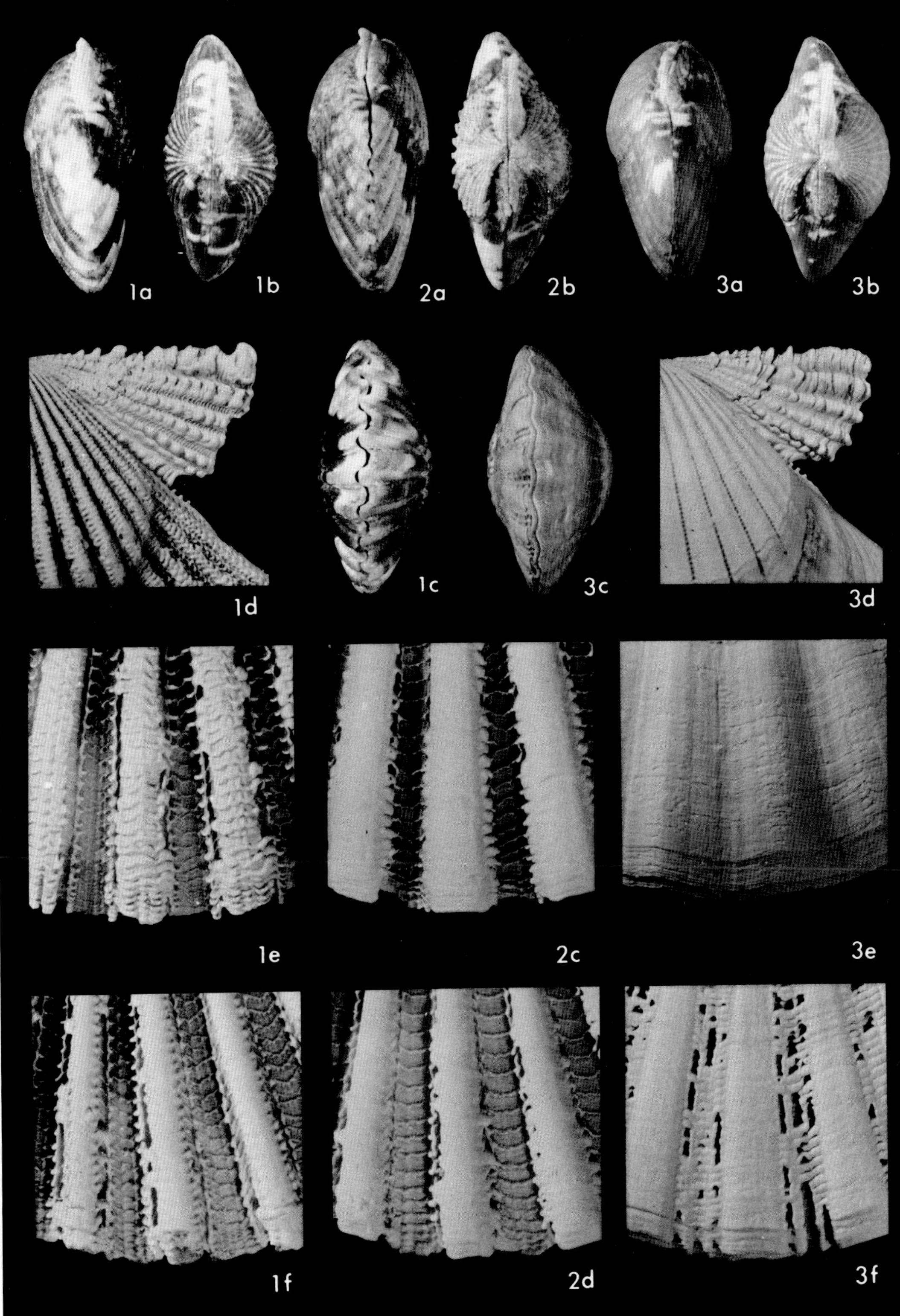

Explanation of Plate 6

(All figures ×1.5)

Figs. 1–6. *Cryptopecten vesiculosus vesiculosus* (Dunker)

Fig. 1.   Right valve (UMUT CM16051a), Phenotype *Q* (*rough* subphenotype), sample *Ab 1*, Late Pleistocene Jizôdô Formation at Atebi of Boso Peninsula; 1a: external view, 1b: dorsal view, 1c: anterior view.

Fig. 2.   Right valve (UMUT CM16051b), Phenotype *Q* (*smooth* subphenotype), the same sample; 2a: external view, 2b: dorsal view, 2c: anterior view.

Fig. 3.   Right valve (UMUT CM16051c), Phenotype *R*, the same sample; 3a: external view, 3b: dorsal view, 3c: anterior view.

Fig. 4.   Left valve (UMUT CM16051d), Phenotype *Q* (*rough* subphenotype), the same sample; 4a: external view, 4b: dorsal view, 4c: anterior view.

Fig. 5.   Left valve (UMUT CM16051e), Phenotype *Q* (*smooth* subphenotype), the same sample; 5a: external view, 5b: dorsal view, 5c: anterior view.

Fig. 6.   Left valve (UMUT CM16051f), Phenotype *R*, the same sample; 6a: external view, 6b: dorsal view, 6c: anterior view.

1a
1b
1c
2a
2b
2c
3a
3b
3c
4a
4b
4c
5a
5b
5c
6a
6b
6c

<h1 style="text-align:center">Explanation of Plate 7</h1>

(All figures ×1.5, unless otherwise stated)

Figs. 1–10. *Cryptopecten vesiculosus vesiculosus* (Dunker)

Fig. 1. Right valve (UMUT CM16057a), Phenotype Q (*rough* subphenotype), sample *Ms*, Holocene Moeshima Shell Bed at Moeshima Islet near Kagoshima.

Fig. 2. Left valve (UMUT CM16057b), Phenotype Q (*rough* subphenotype), the same sample.

Fig. 3. Right valve (UMUT CM16057c), Phenotype Q (*smooth* subphenotype), the same sample.

Fig. 4. Left valve (UMUT CM16057d), Phenotype Q (*smooth* subphenotype), the same sample.

Fig. 5. Right valve (UMUT CM16057e), Phenotype R, the same sample.

Fig. 6. Left valve (UMUT CM16057f), Phenotype R, the same sample.

Fig. 7. Right valve (UMUT CM16038a), Phenotype Q, sample *Tm 2*, Middle Pleistocene Umegase Formation at Tsujimori of Boso Peninsula.

Fig. 8. Left valve (UMUT CM16038b), Phenotype Q, the same sample.

Fig. 9. Conjoined valves (UMUT CM16030a), Phenotype Q, sample *Ik*, Late Pliocene Shinzato Formation at Ikei Islet, Okinawa, 9a: external view of right valve, 9b: external view of left valve, 9c: anterior view, 9d: dorsal view.

Fig. 10. Conjoined valves (UMUT CM16030b), Phenotype Q, the same sample; external view of right valve.

Figs. 11, 12. *Cryptopecten vesiculosus makiyamai* subsp. nov.

Fig. 11. Right valve (UMUT CM16033b), Paratype, sample *Kg 1*, Late Pliocene Hosoya Silt of Kakegawa Group at west of Ugari near Fukuroi. (×1)

Fig. 12. Left valve (UMUT CM16033a), Holotype, the same sample. (×1)

Explanation of Plate 8

(All figures ×1.5, unless otherwise stated)

Figs. 1–5. *Cryptopecten spinosus* sp. nov.

Fig. 1. Right valve (UMUT CM16170d), Paratype, sample *Kk* (*S*), Late Pleistocene Wan Formation at Kamikatetsu, Kikai Island; 1a: external view, 1b: internal view.

Fig. 2. Left valve (UMUT CM16170c), Paratype, the same sample; 2a: external view, 2b: internal view, 2c: dorsal view, 2d: anterior view.

Fig. 3. Right valve (UMUT CM16170e), Holotype, the same sample; 3a: external view, 3b: internal view, 3c: dorsal view, 3d: anterior view, 3e: byssal wing. (×4)

Fig. 4. Left valve (UMUT CM16170f), Paratype, the same sample; 4a: external view, 4b: internal view.

Fig. 5. Ventral area of right valve (UMUT CM16170g), Paratype, the same sample. (×4)

Figs. 6–9. *Cryptopecten yanagawaensis* (Nomura and Zinbo)

Fig. 6. Right valve (UMUT CM16171a), sample *Mn* (*Y*), *Middle* Miocene Moniwa Formation at Junishin near Natori.

Fig. 7. Left valve (UMUT CM16171b), the same sample.

Fig. 8. Left valve (UMUT CM16171c), the same sample.

Fig. 9. Right valve (UMUT CM16171d), the same sample.

Explanation of Plate 9

Fig. 1.   *Cryptopecten bullatus* (Dautzenberg and Bavay)
  Fig. 1.   Living conjoined valves (USNM 173194), Holotype of *Cryptopecten alli* Dall, Bartsch and Rehder, 1938; ×1.5; off south coast of Oahu, Hawaii; 1a: external view of right valve, 1b: internal view of right valve, 1c: external view of left valve, 1d: internal view of left valve.
Figs. 2–5.   *Cryptopecten nux nux* (Reeve)
  Fig. 2.   Living conjoined valves (USNM 764157); ×2; Hervey Bay, Queensland; 2a: external view of right valve, 2b: external view of left valve, 2c: anterior view, 2d: dorsal view.
  Fig. 3.   Living conjoined valves (USNM 764164); ×2; Jolo, Philippines; 3a: external view of right valve, 3b: external view of left valve.
  Fig. 4.   Living conjoined valves (USNM 764160); ×2; Itoman, Okinawa, Japan; 4a: external view of right valve, 4b: external view of left valve.
  Fig. 5.   Right valve (USNM 764165); ×2; Al Ghardaqa, Egypt.
Figs. 6–9.   *Cryptopecten phrygium* (Dall)
  Fig. 6.   Living conjoined valves (USNM 62352); ×1.5; Mexican Gulf; 6a: external view of right valve, 6b: internal view of right valve, 6c: external view of left valve, 6d: internal view of left valve.
  Fig. 7.   Living conjoined valves (USNM 764544); ×2; off Hillsboro, Florida; 7a: external view of right valve, 7b: external view of left valve.
  Fig. 8.   Right valve (USNM 694425); ×1.2; Mexican Gulf.
  Fig. 9.   Left valve (USNM 764542); ×1.2; off southwest Louisiana, Mexican Gulf.

Explanation of Plate 10

(SEM photographs, all figures ×29, 5 KV)

Figs. 1, 2.   *Cryptopecten vesiculosus vesiculosus* (Dunker)

Fig. 1.   Living conjoined valves (UMUT RM16059a), Phenotype *Q*, sample *Hy*, off southwestern coast of Miura Peninsula; 1a: umbonal area of right valve, 1b: umbonal area of left valve.

Fig. 2.   Living conjoined valves (UMUT RM16059b), Phenotype *R*, the same sample; 2a: umbonal area of right valve, 2b: umbonal area of left valve.

Fig. 3.   *Cryptopecten bullatus* (Dautzenberg and Bavay)

Fig. 3.   Living conjoined valves (UMUT RM16002a), sample *Hs* (*3*) (*A*), Hyôtanse near Niijima Island; 3a: umbonal area of right valve, 3b: umbonal area of left valve.

Explanation of Plate 11

(SEM photographs, all figures ×29, 5 KV)

Figs. 1, 2.   *Cryptopecten vesiculosus vesiculosus* (Dunker)

Fig. 1.   Living conjoined valves (UMUT RM16059a), Phenotype Q, the same specimen as Pl. 10, Fig. 1; 1a: middle part of disk of right valve, 1b: middle part of disk of left valve.

Fig. 2.   Living conjoined valves (UMUT RM16059b), Phenotype R, the same specimen as Pl. 10, Fig. 2; 2a: middle part of disk of right valve, 2b: middle part of disk of left valve.

Fig. 3.   *Cryptopecten bullatus* (Dautzenberg and Bavay)

Fig. 3.   Living conjoined valves (UMUT RM16002a), the same specimen as Pl. 10, Fig. 3; 3a: middle part of disk of right valve, 3b: middle part of disk of left valve.

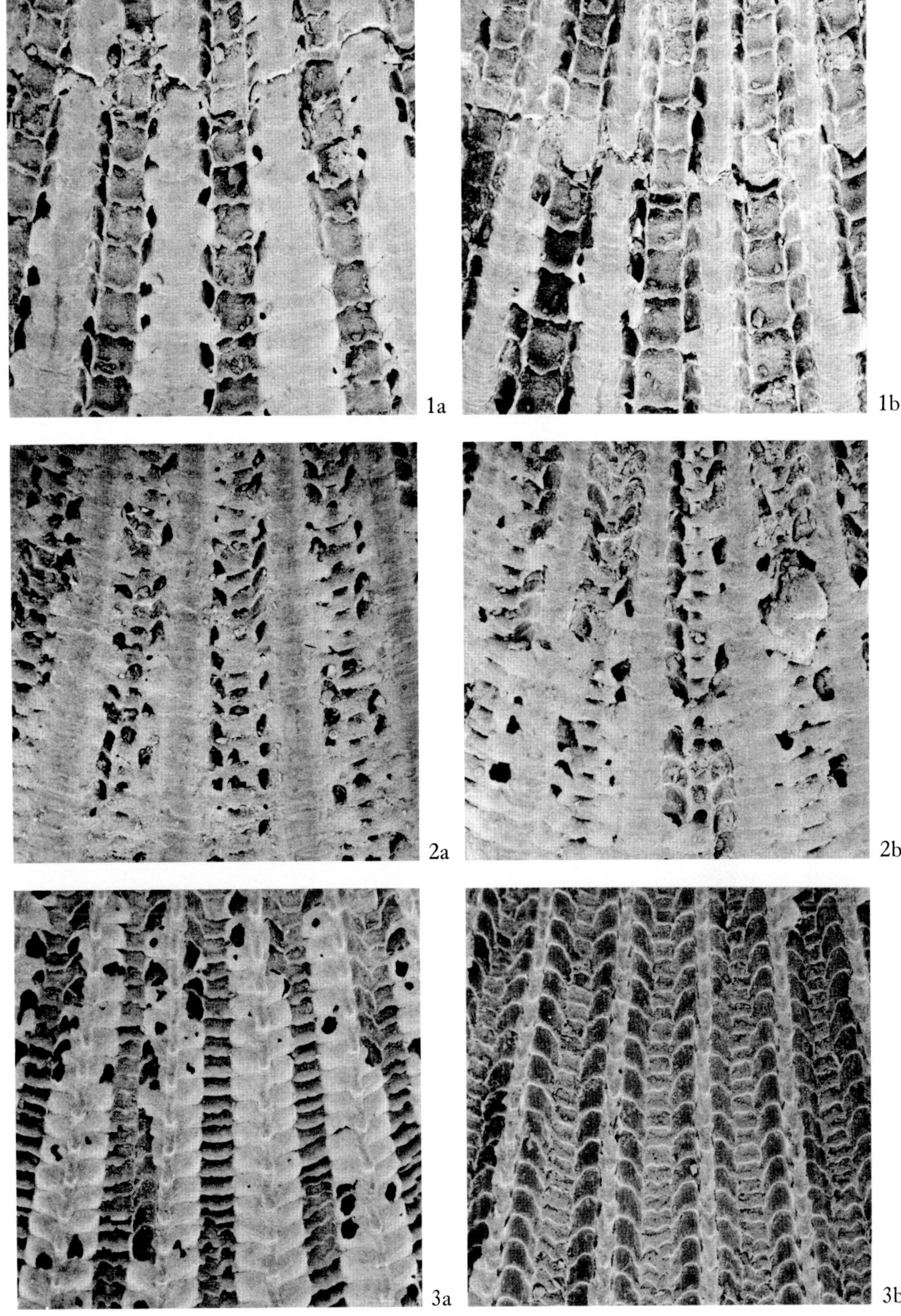

1a
1b
2a
2b
3a
3b

Explanation of Plate 12

(SEM photographs, all figures × 29, 5 KV)

Figs. 1, 2. *Cryptopecten nux nux* (Reeve)

Fig. 1. Umbonal area of right valve (UMUT CM16015a), sample *Kk* (*N*), Late Pleistocene Wan Formation at Kikai Island.

Fig. 2. Umbonal area of left valve (UMUT CM16015b), the same sample.

Figs. 3–5. *Cryptopecten spinosus* sp. nov.

Fig. 3. Umbonal area of right valve (UMUT CM16170a), sample *Kk* (*S*), Late Pleistocene Wan Formation at Kikai Island.

Fig. 4. Umbonal area of left valve (UMUT CM16170b), the same sample.

Fig. 5. Radial ribs on ventral area of right valve (UMUT CM16170a), the same specimen as Pl. 12, Fig. 3, showing oppositely disposed imbricated scales.

Fig. 6. *Cryptopecten vesiculosus vesiculosus* (Dunker)

Fig. 6. Radial rib on middle-ventral area of adult left valve (UMUT RM16061a), Phenotype *R*, sample *Jg* (*2*), off Jôgashima Islet, Sagami Bay, showing oppositely disposed imbricated scales.

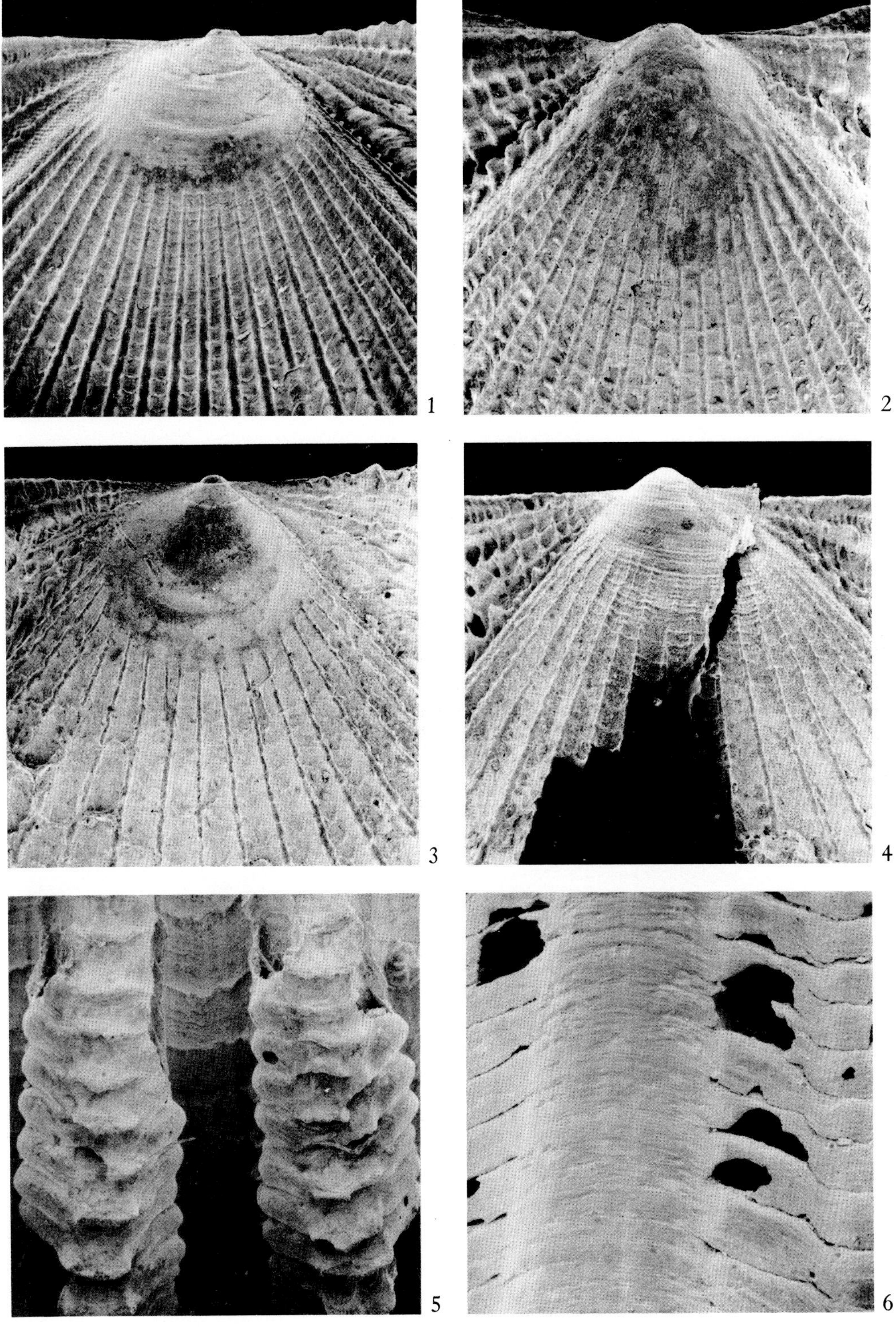

Explanation of Plate 13

(SEM photographs, all figures ×29, 5 KV)

Figs. 1, 2. *Cryptopecten vesiculosus vesiculosus* (Dunker)

Fig. 1. Living conjoined valves (UMUT RM16138a), Phenotype *Q*, sample *Is*, mouth of Ise Bay; 1a: transverse section of radial rib at middle part of disk of right valve; 1b: oblique view of interspace of radial ribs at ventral margin of right valve, 1c: transverse section of radial rib at middle part of disk of left valve; 1d: oblique view of interspace of radial ribs at ventral margin of left valve.

Fig. 2. Living conjoined valves (UMUT RM16061a), Phenotype *R*, sample *Jg* (*2*), off Jôgashima Islet, Sagami Bay; 2a: transverse section of radial rib at middle part of disk of right valve, 2b: oblique view of interspace of radial ribs at ventral margin of right valve, 2c: transverse section of radial rib at middle part of disk of left valve, 2d: oblique view of interspace of radial ribs at ventral margin of left valve.

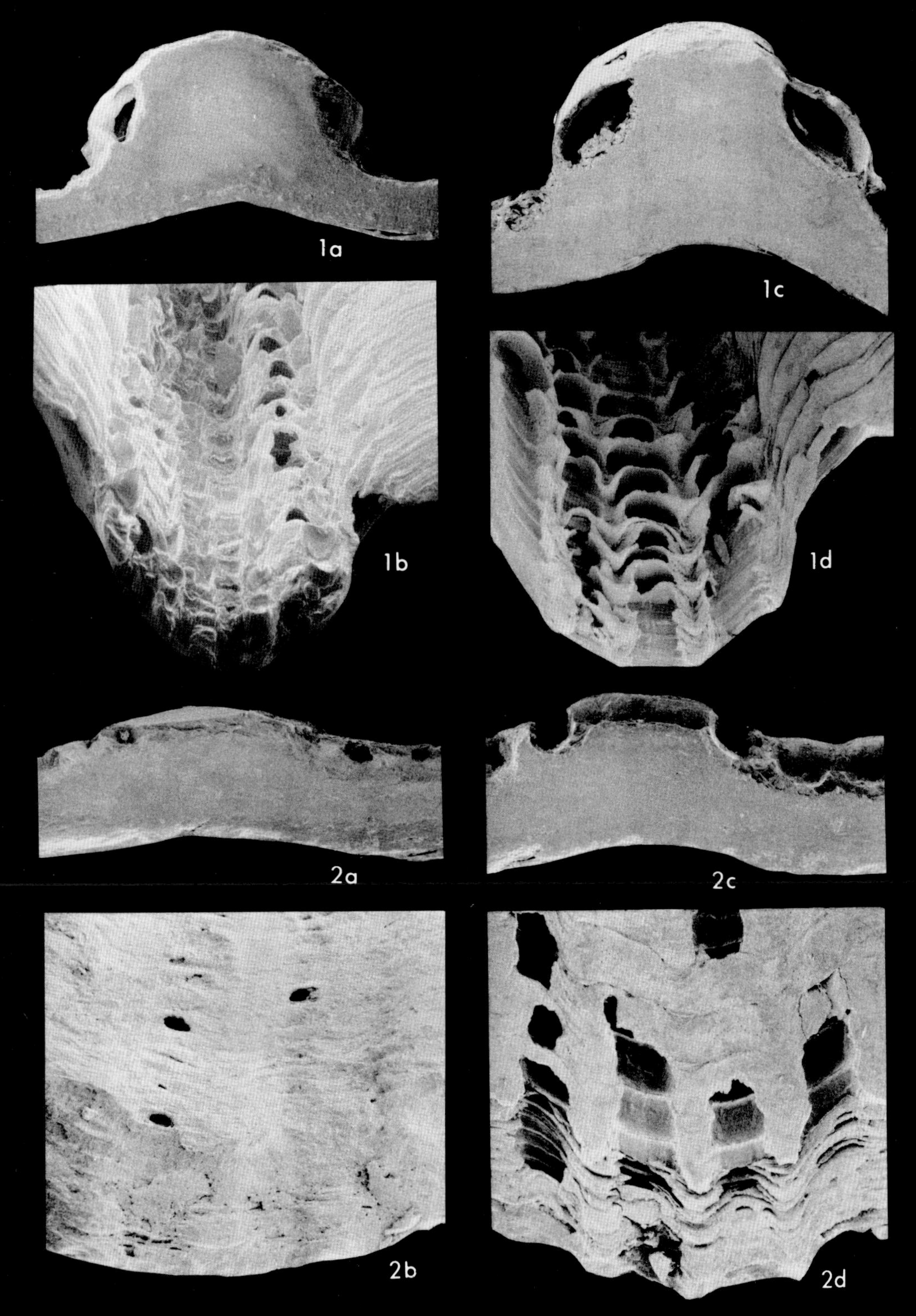